M( )SION PRED. TION ON VEHICLES (1950-PRESENT DAY)

Your Guide to:

Restoring Classic Cars.
Buying Wisely Second Hand.
Motor Vehicle Corrosion Warranties and Their Problems.
Claiming Successfully on the 6 year Motor Vehicle Corrosion Warranty.

by
Hugh McArthur
Ph D. M A (Cantab). M Inst Corr Tech.

Author's Note: The material published in this title has been prepared for the general information of the reader and while the material is believed to be technically correct, the author does not represent or warrant its suitability for any general or specific use without first securing competent advice. Furthermore, most warranties will not allow you to add to the manufacturer's treatment, or make any alterations to the vehicle. As corrosion is design orientated, and many vehicles are not designed to resist corrosion, the author cannot assume, in this respect, liability or responsibility of any kind in connection with the information herein which is used without first securing competent advice as to the effect upon the warranty (eg. plates 3, 4, 6, 10, 11, 12, 17 and 18).

Hugh McArthur Ph D. M A (Cantab). M Inst Corr Tech.
Expert Books
Woodhouse Eaves. 18 LE12 8SD

British Library Cataloguing in Publication Data
McArthur, Hugh
MOTOR VEHICLE CORROSION, PREDICTION AND PREVENTION

Your Guide to:
Restoring Classic Cars.
Buying Wisely Second Hand.
Motor Vehicle Corrosion Warranties and Their Problems.
Claiming Successfully on the 6 year Motor Vehicle Corrosion Warranty.
1 Motor Vehicles. Corrosion.
I Title
629.232

ISBN 0 9515787 0 7 Paperback
ISBN 0 9515787 1 5 Hardback

First published in 1990 by:
Expert Books.
Woodhouse Eaves. LE12 8SD
Tel: 0509 890607

Printed in Great Britain by: Echo Press (1983) Ltd., Echo House, Belton Park, Jubilee Drive, Loughborough, Leicestershire.

Previous Publications.

(1) McArthur H. Autocar October 1973.
(2) McArthur H. 1979 Engineering Oct.1308-15.
(3) McArthur H. International Corrosion Conference Adelaide Australia 1980.
(4) McArthur H. 1980 Inst. Mech. Engnrs. Conference "Impact of Design on Whole Life Costing Paper No.359/80 LONDON.
(5) McArthur H. Automotive Engineer 1981. Oct/Nov. Vol.6, No.5, 18-21
(6) McArthur H. Jnl. Oil. & Colour Chemists 1982. Aug. p.306-316.
(7) McArthur H. Jnl. of Metals 1985 Jan. V1. No 1 p. 28-32.
(8) McArthur H. International Conference Toronto Canada 1984.
(9) McArthur H. Jnl. Metals & Materials 1985 Jan.
(10) McArthur H. International Corrosion Conference Chicago Illinois 1985.
(11) McArthur H. International Corrosion Conference Tokyo Japan 1985
(12) McArthur H. International Corrosion Conference Melbourne Australia 1987.
(13) McArthur H. Corrosion Prediction and Prevention in Motor Vehicles. Ellis Horwood, Chichester and Halstead Press, ISBN 0-7458-0357-1 Ellis Horwood, and ISBN 0-470-20995-X Halstead Press].

Typeset in Plantin Roman by
Expert Books (Tel. 0509 890607) and Echo Press (1983) Ltd., Loughborough

"Sorry Sir! It's Stone Chipping."

How to Claim Successfully on the 6 year Motor Vehicle Corrosion Warranty

Contents:

Preface.

The idea for this guide came from the following: (**a**) the motor vehicle designers who wanted a thumbnail sketch of the corrosion process with respect to motor cars, and how one side zinc coated steels would behave in their motor cars; (**b**) members of the public wanted to know why zinc could not be placed beneath the wheel arch, and could, for example, zinc which made electrical contact with the car offer sacrificial protection if it were placed in the windscreen washer bottle to corrode; (**c**) the general public who wanted help on their corrosion warranties; and (**d**) the Ministry of Transport (Corrosion Review Committee), who wanted to know where to direct the nominated testers to find corrosion on a wide range of models and makes.

Lord MacMillan once said of arbitration with expert witnesses: "I am not certain that any scientific man ought even to become partisan of an opinion, but ought never to accept a retainer to advocate a particular view merely because it is the view which is in the interests of the party who has retained him to maintain. To do so is to prostitute science and to practice a fraud on the administration of justice. The trust role of an expert witness is to offer the Court the ***best assistance*** he can, by getting at the truth". I hope that, by writing this book, I have given the ***best assistance*** to both the layman and the courts, and moreover given an ***unbiased explanation*** of corrosion and its relationship to ***design*** which benefits motorists and manufacturers alike.

Glossary of Terms.

Acid and alkaline are terms used to express the acidity of a solution. Steel dissolves ***rapidly*** in acidic mediums, but is ***chemically protected*** in alkaline mediums.

Anode is the negative terminal of an elemental battery. It is the region which dissolves in the corrosion process.

Capillary wicking. Capillarity is a phenomenon caused by surface tension resulting in a distortion (raising or lowering) of the surface of a liquid from its true equilibrium (figure 21a).

Cathode is the positive terminal of an elemental battery. It is the region which does not dissolve in the corrosion process. The chemical by-products produced at the cathode by the flow of the ionic (figure 4) electrical current are alkaline and act as a good paint stripper.

Cathodic protection (1.5, 7.6) is a technique used for protecting steel structures from electrolytic corrosion. Cathodic protection effectively reverses the natural corrosion current by using an externally applied current and voltage. The externally applied process voltage is insufficient to plate the metallic ions out as the metal.

Current. See electron.

Electrolyte. (1.4, figure 4, 2.0). Batteries have to contain an electrolyte through which the driven current must pass. The 12V car battery contains sulphuric acid and distilled water. If the sulphuric acid were not present, the high currents needed to start the car could not be produced.

Electron. A stable atomic particle having mass and a negative charge. The electron is responsible for the flow of an electrical current in a metallic conductor. For the current to flow in a liquid the electron must be changed to an ***ion.*** The electrical current flowing in a liquid can be composed of the movement of both positive and negative ions. Positive charge cannot move in a metallic conductor.

Holidays (in paints and mastics) refers to small breaks in the otherwise adherent and coherent continuous coating.

Ions are produced when a substance dissolves in water (1.4), increasing the conductivity of water thus producing an electrolyte. The current carried within the electrolyte is carried by these ions and is called the ionic current.

Lemon laws (lemon) is an American slang word which means the person or thing is considered useless or defective. I have used the phrase ***bad design*** (1.11) instead. This is referring to bad designs in the corrosion sense, and should not be confused with aesthetic designs. The Americans have laws (***lemon laws***) to protect the consumer against defective goods. In the trade a ***lemon car*** is one which has been assembled with a high proportion of faults which have passed the inspection process and not rectified by the pre-delivery inspection (a Friday night car).

Micron a unit of length equal to 10^{-9} of a metre.

Momentum is the product of mass and velocity.

GLOSSARY continued on page 22

CHAPTER 1

The Corrosion Reaction as a Battery

1.0 INTRODUCTION

Many private motorists are concerned with the deterioration of their motor vehicle by corrosion. Important questions arise:

1) Why do the panels rot through from the inside to the outside? (See plates 3, 6, 7, 9, 11, 12, 13, and 19).
2) Why have cars traditionally suffered from this form of corrosion?. (Chapter 8.10)
3) How can you detect and differentiate between warranty excluded 'scab' corrosion, which arises as a result of stone chipping and the very similar looking paint blister which accompanies metal perforation which comes under the corrosion warranty?. (Chapter 9.0)
4) Are the manufacturer's agents fair when they claim that the rust is due to stone chipping?, (eg plate 1)
5) What is the 6/8/10 year corrosion warranty and how can you claim?. To make a successful claim, you will have to know equally as much as the dealers, recognise the merits (and faults) of alternative approaches, be fair, have patience, be persistent, and know the correct answer to counteract the often heard phrase: **"Sorry Sir! It's stone chipping!"**, eg see plates 1, 6, 8, 9, 15. (One manufacturer was so good at convincing owners that the corrosion around the rear wheel arch and doors was due to stone chipping that they were able to introduce the replacement model with plastic guards to prevent car park knocks and stone chips from damaging the paint. The real culprit was due to water ingress into the doors by badly designed glazing rubbers (plate 16) and the passenger air extraction system allowing melting snow to enter the rear wheel arch area (Chapter 5.17).
6) What items are excluded from the warranty? What items are included in the warranty? (See chapter 8.0-8.8)
7) What should the replies be to the warranty agent or garage in any dispute?. Be placid, realistic, and negotiate by having a superior technical knowledge which this book will generate.
8) Can the rust be "killed"? (Chapters 1.3, 10.19).
9) How do mud pockets form under the wheel arch when there is so much road spray? (Chapter 5.1 to 5.8).
10) Couldn't this road spray wash the wings and their wheel arch lip returns? (Chapter 5.3 to 5.5).
11) What causes stone chips on the doors? (Chapter 5.1).

12) How can you predict the sites of motor vehicle corrosion and where should the motorist look for the telltale signs of corrosion in nearly new and used cars? (Chapter 10 Summary)
13) Is it possible to make a better judgment of the state of corrosion in a used car ? (eg Plates 6 and 18).
14) Can the MOT test be relied upon as a guarantee of road worthiness when purchasing a vehicle with a "new 12 month" MOT?. (Chapters 3.3, 9.1 and sequence)
15) Will the MOT test detect corrosion in panels which are subject to a 6 year warranty? (Chapters 3.2, 5.7(e), 9.0 and sequence).
16) Should the manufacturer's goodwill gesture of 50% towards the cost of repair be accepted? (Chapter 10.18 and sequence)
17) Should you purchase a car with a lapsed 6 year warranty? (Chapter 8.4).

1.1 FURTHER READING

This book describes the mechanisms which take place to produce internally nucleated corrosion within the car body. Owing to a shortage of space, other technical aspects of corrosion to which reference is made by the symbol (•), can be followed up from the author's book titled: ***Corrosion Prediction and Prevention in the Motor Vehicle*** (published by Ellis Horwood, Chichester and Halstead Press, under the following ISBN numbers: ISBN 0-7458-0357-1 Ellis Horwood, and ISBN 0-470-20995-X Halstead Press).

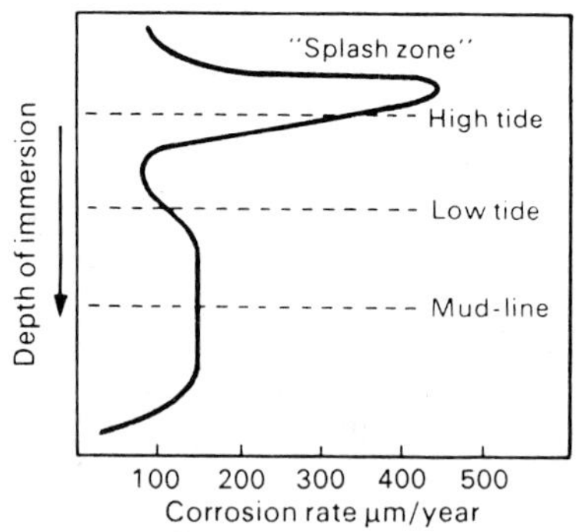

FIGURE 1. Corrosion rates of steel for different positions in tidal sea water. Note that the splash zone has a higher corrosion rate than steel immersed at high tide. This higher rate of corrosion in sea spray results from the sea spray being well oxygenated, (figure 5, (1.8)). During the winter, the road spray is essentially recycled sea water spray. ***All*** vehicles on the road during the winter are subjected to this corrosive "sea spray" and this fact should be appreciated by every motorist. Temperature also affects the corrosion rate. Rough rules are: (1) A temperature rise of 10C results in a doubling of the corrosion rate. (2) A temperature rise will dry the salt droplet quickly unless prevented by external means (figures 3 and 22), and therefore could reduce the corrosion rate (2.0(a)). Figure contributed by the Department of Trade and Industry and acknowledged with thanks.

1.2 CROSS REFERENCING

Corrosion embraces the study of metals, physics and chemistry. These are just the technical subjects which most people give up at school because they find them too hard. Therefore to try and explain corrosion to the layman is going to be difficult. To help you through the subject, I have numbered each paragraph. Each paragraph contains a ***principle*** which is required as basic knowledge to understand corrosion. Cross reference to other sections is shown by numbers in brackets, eg (1.17).

For readers who may not have read technical books before, and those new to corrosion, I suggest that you read the summary section (10.0) first to get a quick review. Omit the cross references, (but refer to the plates and figures). As your knowledge increases, take up some of the cross references in chapter 10. As you feel that your knowledge is increasing start at chapter 1 and proceed through the text slowly. Also for your convenience, chapters 6 and 7 are summarised at the end of each chapter. A glossary is given for those technical terms which I felt had to be used. Please don't make the mistake of believing that you should understand these principles at a first reading, because it will not happen this way. Science is a strict discipline even for the qualified engineer (8.10(7)).

PRINCIPLES

1.3 WHAT IS CORROSION?

The word "corrosion" is derived from the Latin word corrodere which means to gnaw away, or to wear away. Indeed the dictionary definition is: 'a process in which a solid, especially a metal, is eaten away and changed by a chemical action, in the presence of water by an electrolytic process'. Corrosion is not an organic pest, or fungus like dry rot in timber – the dry rot of steel. It ***can-not*** be killed. Once the metal is converted into its oxide, ***rust***, the metal cannot be 'regenerated' from the rust by any 'magic' liquid potion. The oxide, rust, is very brittle, and has little energy absorbing properties (●) (3.5). Rust also contains water and harbours road salts, thus aggravating the problem of corrosion.

1.4 CORROSION AS A BATTERY**

The corrosion process on a car body can be considered as an ***unwanted*** and ***unexpected 'battery'***. A battery is an electrolytic process which is why the term "electrolytic process" occurs in most dictionary definitions of corrosion. The concept of corrosion as a ***battery*** will be a new concept to many and so it is worth investigating further.

First of all we need to know what is meant by the term 'electrolytic process'. Most readers will be conversant with the electroplating of metals.

In this publication we will refer to the corrosion ***cell* as a ***battery*** to remind us that the concept of corrosion is a short circuited battery with ***no*** switch (10.2).

In order to plate a metal, the dissolved salts of the metal must be soluble in water. We call the solution formed an electrolyte and the metal is dissolved as charged particles which we call ***ions***. The action of plating the metal is achieved by applying an electric current, whose function is to change the dissolved ions into the plated metal. This can only take place on the metal and at the metal/liquid interface. The process by which this takes place is an 'electrolytic process'. A battery ***produces*** a current by the reverse process, of a metal going into solution to form ions by another 'electrolytic process'. The electrical current flowing in the electrolyte is referred to as the ***ionic current*** to differentiate it from the current (of electrons) which flows in the metallic conductors.

1.5 ELECTROLYTIC REACTIONS

Examples of electrolytic reactions considered in this book are:

Electrical battery: A metal dissolves producing ions and a direct current and voltage. The corrosion cell or battery gives a short circuited voltage (figure 6d).

*Electro**plating:*** Dissolved metallic ions deposited by an externally applied direct (***impressed***) current, (rather like a reverse battery process).

*Electro**coating*** (4.3): Water borne paint deposited by an externally applied direct current. The process can be anodic (old system •) or cathodic (new system) and is often called cathodic electrophoresis or cataphoresis in advertisements. The process is the paint equivalent of the electrodeposition of a metal.

Cathodic protection (7.3): An electrical process which protects steel by making the steel structure the cathode of the cell either by an externally applied direct (***impressed***) current or by connecting the structure to a more active metal eg zinc. The electrical connections are similar to ***electroplating*** but any dissolved metallic ions are ***not deposited.***

Corrosion. Corrosion is the result of a chemical ***electrolytic*** reaction brought about by:

a) a metal being wet. The water can be produced by the condensation of atmospheric water vapour to form a liquid. The metal does not have to be ***visibly*** wet;

b) the wetness is accompanied by a difference in the oxygen replenishment at the steel surface beneath the wetted area (figure 1). This will be further developed in the following sections so that the water droplet, (or sea spray), sits on the steel surface with a hemispherical profile as shown in figure 6b. The perimeter of the droplet allows easy oxygen replenishment to the metal surface compared to the central area which is relatively depleted in oxygen.

Both these conditions of differential aeration and wetness are most easily achieved with a "**crevice**" (figures 2, and 3,).

1.6 AREAS OF NATURAL WETNESS. CREVICES

Crevices trap water for long periods of time and concentrate the road de-icing salts within them (6.3). The presence of the road de-icing salts within a crevice allows the crevice to ***attract*** moisture vapour from the air and form a moist environment, (for example during a winter in the UK, some of these crevices would never dry). A crevice also has the ***natural*** ability to provide the difference in the oxygen replenishment at the air/steel, (or steel/paint interface), which is the main requirement to drive the 'corrosion battery'.

These crevices can be created by:

1) a metal overlap; a spot weld; a bolted section;
2) two adjacent pieces of metal, figure 3;
3) metal perpetually in contact with plastic, eg rear wheel arch stone guards;
4) or around a dirt particle which rests ***perpetually*** in the ***same place*** upon a metal surface (figure 6);
5) beneath mud accumulations (figure 18);
6) beneath non adherent paint films, and at the start of ***'under-cutting'*** (figures 8 and 9, (2.6)).

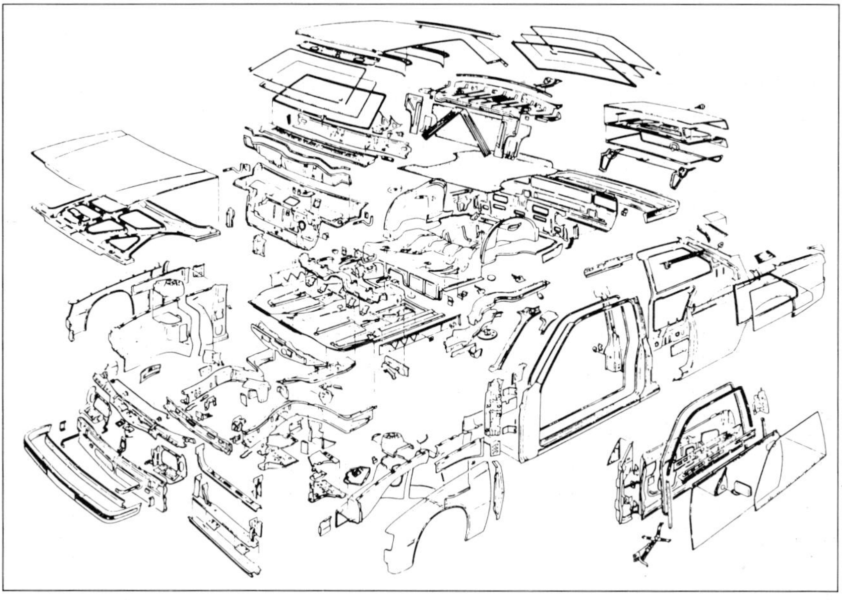

FIGURE 2. Exploded view of the panels making up a vehicle body. Figure contributed by Vauxhall Motors, Luton by courtesy of General Motors.

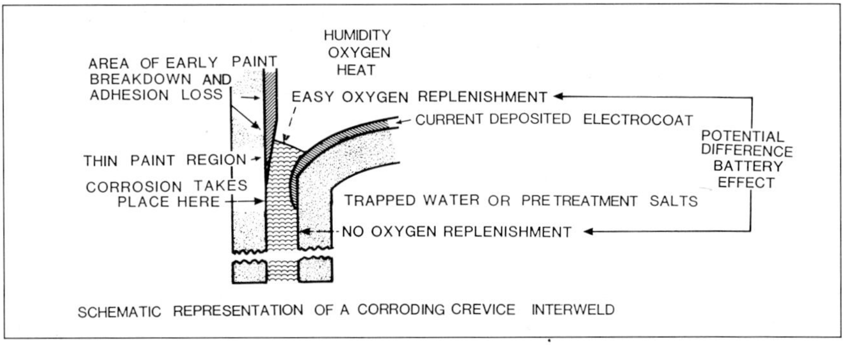

FIGURE 3. Shows the lack of paint thickness in the crevice. A water filled crevice makes an ideal corrosion cell. The external surface of the water envelope is well oxygenated, whilst the narrowest part of the crevice contains stagnant, deoxygenated water. The metal in contact with this deoxygenated water will corrode, whilst the oxygenated area produces a chemical by-product (1.10) which will chemically strip the paint from the metal substrate. This loss of adhesion causes the removal of the protective paint and wax films (2.6).

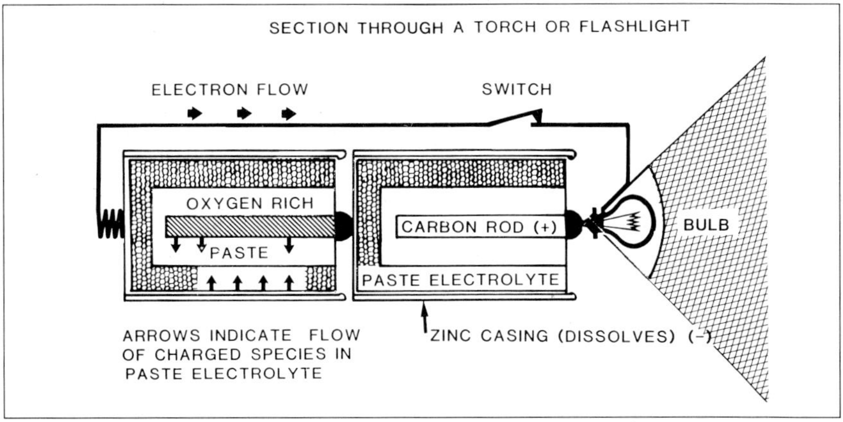

FIGURE 4. A cross section through a typical torch, complete with batteries. A battery can be considered as being made up of two ***half cells***. Both half cells make up a cell or battery. One half cell is the ***anode*** or (-ve) electrode at which certain chemical reactions take place, (called anodic reactions), and the other half cell is the ***cathode*** or (+ve) electrode. These two half cells are in a 'paste electrolyte'. The electrical current, (ie the electrons), are shown flowing in the ***metallic*** external circuit, and the charged (**ionic**) species from both half cells combine in the "paste electrolyte" to produce a new compound and provide the complete electrical circuit. The electrical energy is produced by the zinc case dissolving, called the anodic reaction. The agency producing the electric current (the ***electrons***) is consumed at the carbon rod by a cathodic reaction. The torch is switched off by stopping the flow of electrons in the outer metallic circuit reaching the cathode. In the dry cell battery, the corrosion process, (which produces the battery voltage), has been made more efficient by research and development. When the metal has been consumed, the voltage drops and the battery is thrown away, (hopefully before it leaks!). Proceed to figure 5.

1.7 COMMERCIAL BATTERIES AS A CORROSION PROCESS

The 1.5V battery which we use in torches and transistor radios, are called dry cell batteries and produce an electrical current by a corrosion process, (figure 4).

1.8 THE CORROSION BATTERY. REDUCING CORROSION

The corrosion taking place on a motor vehicle obtains its current from the differences in the replenishment of oxygen at the steel surface which plays host to both anodic and cathodic (half cell) reactions. It could be described as an ***'oxygen in water charged'*** battery, or ***differential aeration cell***, (figure 5). Such a short circuited battery is shown in figure 6d. The wetted steel surface which is ***perpetually*** more denuded of oxygen is the area which corrodes.

The commercial 1.5V dry cell battery can be switched off by disconnecting the outside metallic part of the circuit but there is no switch to switch off the 'equivalent corrosion battery'. We can only stop the 'corrosion battery' by removing the liquid environment between the anode and cathode sites. This is equivalent to throwing away the electrolyte in your well charged 12V car battery. If you did this to your car battery, there would be no life in the battery at all.

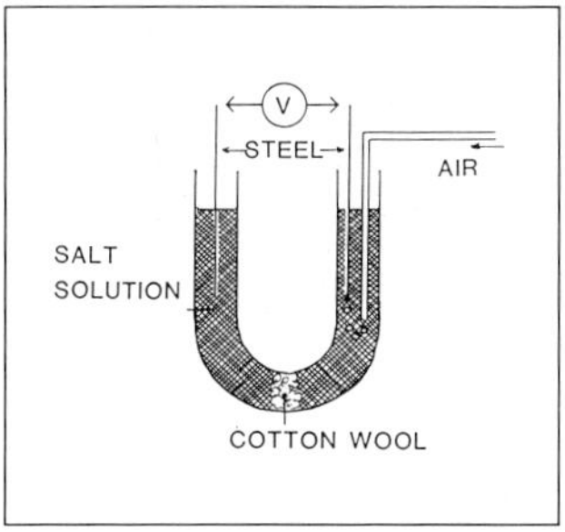

FIGURE 5. This shows how you could demonstrate the two half cells associated with the 'corrosion battery'. Two pieces of mild steel are connected to a voltmeter (V) as shown, and immersed in a saline solution (the ***electrolyte***) which is differentially oxygenated to produce a voltage difference. The voltage difference developed increases with increased oxygenation of one arm (ie one half cell), with respect to the adjacent arm or half cell. If the two arms are oxygenated to the same extent, there is **no** voltage difference and in this state can be likened to a 'run down' battery.

There is ***no*** 'corrosion current' and therefore no corrosion if:

a) there is ***no*** difference in oxygen replenishment rates on the metal surface beneath the area of wetness; (eg metal artifacts from the Marie Rose buried in the mud at the bottom of the sea, or stone chips on the bonnet, figure 6f).

b) anodic sites are changing into cathodic sites and vice versa, (eg rain water on the painted car roof. This is discussed in (1.10, and 2.8)).

c) the product of the anodic reaction cannot combine with the products of the cathodic reaction, (eg in the application of sacrificial anodes, plate 20, (or similar lumps of zinc corroding either in the rear wheel arch, or in the windscreen washer bottle, (discussed in 7.5)).

If I were to summarise the requirements to eliminate or reduce corrosion, it would have to be to keep the vehicle crevices as dry as possible, and to have equal oxygen availability to the whole steel surface. This is basically what a ***clean*** paint film achieves. This intent is not achieved by the designs photographed to illustrate corrosion in this booklet, (eg. see Plates 1, 3, 4, 6-19), because all protective films break down in contact with water over long periods of time. Therefore, it is important to avoid water and particulate traps within the vehicle body (eg plates 3, 4, 6, 7, 10, 11, 12, 13, 14, 15, 17, and 18).

1.9 ROAD DIRT AS BATTERIES

The series of diagrams, (figure 6), shows what happens inside the box members.

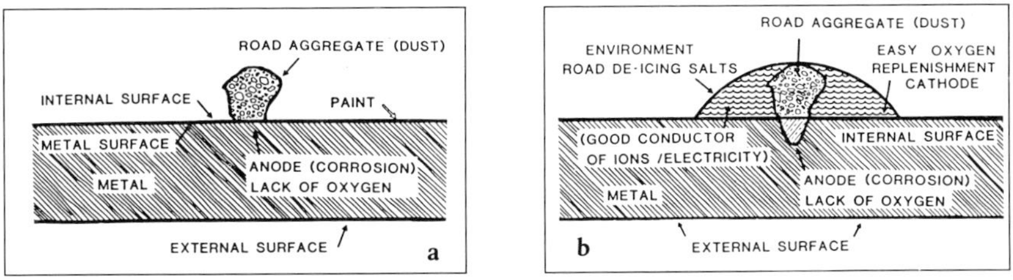

Figure 6 (a) view of the road dirt or dust contaminated with the winter road de-icing salts resting on the surface of the car body.

Figure 6 (b) view of the road de-icing salts absorbing water from the atmosphere to form the differential aeration cell. (The water envelope is exaggerated for schematic purposes. In practice only the dirt/paint interface would be wet).

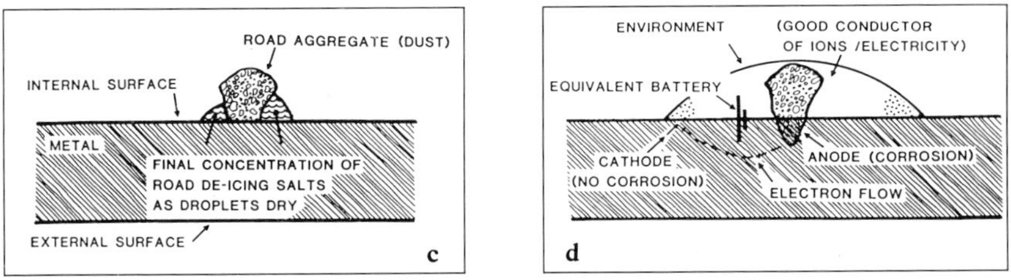

Figure 6 (c) road de-icing salts concentrating at the base of the road dirt upon drying which 'adhesively' fixes the aggressive environment upon vertical surfaces. Each time the humidity increases, the salt will become moist ensuring that the anodic sites are always anodic sites and cathodic sites are always cathodic sites, thereby ensuring a ready made corrosion battery when the liquid environment is restored.

Figure 6 (d) The short circuited equivalent battery is generated by the difference in oxygen availability. The well oxygenated region is the perimeter of the water droplet and is the ***cathode*** or (+ve) electrode of the 'corrosion battery'. The deoxygenated region, and the area which dissolves is the centre underneath the water droplet, and is the ***anode*** or (−ve) electrode of the 'corrosion battery'. The anode area is usually very small compared to the cathodic area. This ***large cathode/small anode*** accentuates the metal ***perforation*** problem (figure 7). Compare this state to the state of the paint chip due to a stone (f).

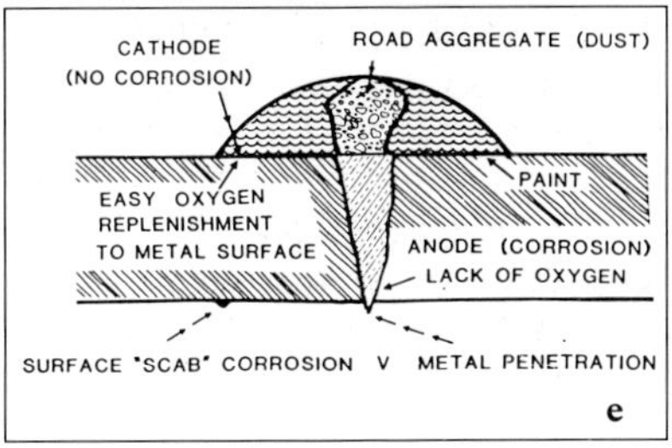

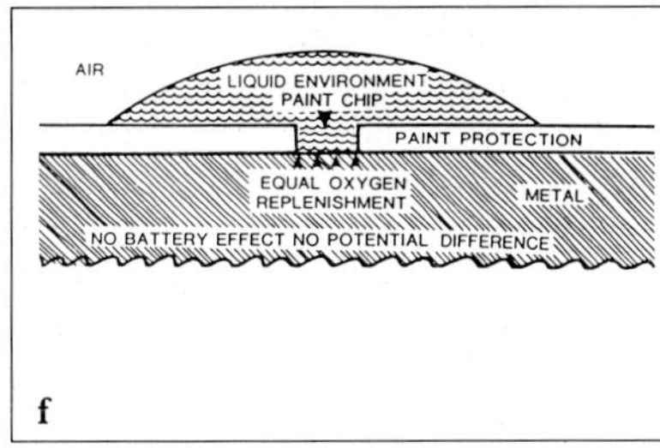

Figure 6 (e) Diagrammatically shows the difference between internally initiated corrosion leading to metal perforation due to the lack of oxygen replenishment beneath these surface deposits, (see also figure 25), and surface initiated corrosion, (known as ***scab corrosion***, figure 9), resulting from stone chips.

Figure 6 (f) A stone chip does not cause very severe corrosion in the early stages of life because:

a) the exposed metal surface is equally oxygenated and
b) the paint chip dries rapidly.

Anodic and cathodic areas do not enter into the problem as this is effectively a 'run down' corrosion cell by result of (a) above.

1.10 BY-PRODUCTS OF CORROSION

In order to explain why paints strip off the substrate metal as a result of corrosion, and why once started, metal perforation is fairly rapid, you will have to take on trust (●) that the chemical by-products of the electrolytic corrosion reaction produce:

1) an acidic environment at the anodic sites (figure 7) where the metal is dissolving. The acidic environment accelerates the corrosion of the metal, thereby accelerating the perforation of the metal. If this acidic environment were not produced, the panel would not perforate within 6 years;
2) an alkaline environment at the cathodic sites. The alkaline environment destroys the adherence of the paint film to the metal substrate, producing a paint blister, termed scab corrosion, (plate 1). ***NO decorative and protective film, (even if it cost as much as gold), will withstand the long term contact effects of adjacent alkaline and acidic environments which are produced by the corrosion reaction in neutral water*** (plate 1).

(An easier concept is to consider water made up of two parts, an acidic part and an alkaline part. These are present in equal amounts and combine together to produce neutral water (H_2O). Effectively the corrosion reaction is isolating these acidic and alkaline parts to the anodic and cathodic sites respectively. Steel dissolves ***rapidly*** in acidic mediums, but is ***chemically protected*** in alkaline mediums. The acid/alkali separation is only established at perpetual anodic and cathodic sites. Hence sites which are changing from anodic to cathodic are ***chemically prevented from corroding. The actual mechanism is explained elsewhere.*** (●)

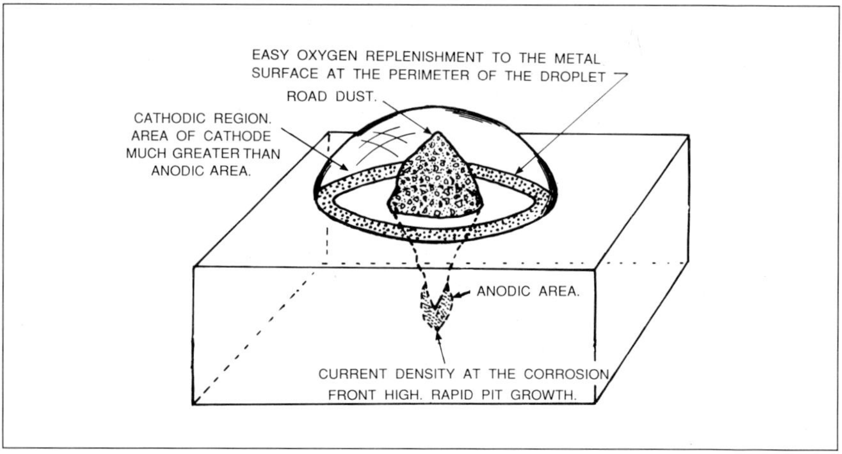

FIGURE 7. Readers who are conversant with current densities (current/area) will appreciate that the anodic current density will be much higher than the cathodic current density due to the much larger external cathodic perimeter. This ***large*** cathode/***small*** anode leads to pitting of the metal. You should also appreciate that as the corrosion front penetrates the metal, the oxygen denuded 'battery' becomes more efficient and the pit becomes an acidic environment. Hence your corrosion warranty becomes more 'valuable' in the last year before expiry. Figure 7 drawn by Maurice Neal and acknowledged with thanks.

This aggressive acidic environment is produced at crevices (1.6). The corrosion process naturally makes a ***perpetually*** oxygen denuded area acidic in nature thus promoting the corrosion reaction (figures 3, 7, 9 and 10). Where panels contain holes for aerials, key holes, washer nozzles etc, this aggressive environment and associated paint stripping will occur on both sides of the wetted crevice. Therefore DIY and garage rectification must be from both sides of the panel.

1.11 CORROSION AND BAD DESIGN

Corrosion has been described as an unwanted, ***unexpected*** natural battery. There are several corrosion sites on a car. Most corrosion sites are inadvertently produced by oversights in design, (ie they are 'inadvertent bad designs', the Americans referring to these items as 'lemons' meaning a dud one). Bad design is used here in the corrosion sense and is defined as any design which allows the accumulation of a corrosive micro-climate, of which most people are unaware, but which the motor manufacturers should be aware of in view of their expertise in the subject matter. Therefore, it will be an advantage for you, to understand the nature of, and inter-relationship between:

1) the differential aeration corrosion mechanism applied to 'crevices', and especially to particles of dirt contaminated by road de-icing salts, which become ***perpetually*** fixed in the same place as a lifetime free rider (figure 6c)
2) the vehicle build and corrosion (figure 2, (3.0)). Each spot weld is a crevice, (figure 3);
3) corrosion and the mass production painting processes, (4.0 to 4.5);
4) corrosion as related to design;
5) vehicle motion and kinetics, (5.0 to 5.14). The effect of vehicle motion on a wet motorway, where de-icing salts are used regularly both to clear snow and prevent the formation of black ice on frosty nights, is very important;
6) the terms and conditions of the manufacturer's or 'after market' warranty (8.0 to 8.10).
7) the materials used by the manufacturers to market the 6/8/10 year warranty (6.0, 7.0).

1.12 SUMMARY

See chapter 10.2.

CHAPTER 2

Corrosion Protection by Paints

2.0 CORROSION PREVENTION

Since I have defined the corrosion of the metal as an 'unwanted or unexpected' short circuited battery brought about by bad design, there is only way of stopping it, and that is by ***good design***. Corrosion can only be 'designed out' by:

a) removing the liquid part* of the 'corrosion battery,' ie between the anode site (−ve electrode) and cathode site (+ve electrode).

* If you threw away the electrolyte in your well charged 12V car battery, there would be ***no life*** in the battery at all, and this would be irrespective of how well charged it was beforehand;

b) preventing the conditions which favour differential oxygen replenishment, now called differential aeration.

2.1 CONTROLLING CORROSION

It is often impracticable to ***eliminate*** corrosion, but we can take steps to reduce or control it. One method of controlling the corrosion process is to make the passage of the electrical current, associated with the chemical by-products (1.10), as small as possible. This is achieved by ***keying*** a paint film onto the surface of the metal (figure, 8) thereby reducing the ***ionic current*** (1.4, figure 4). If the ionic current is reduced, corrosion is reduced. Surface preparation, ie wire brushing and abrading the surface

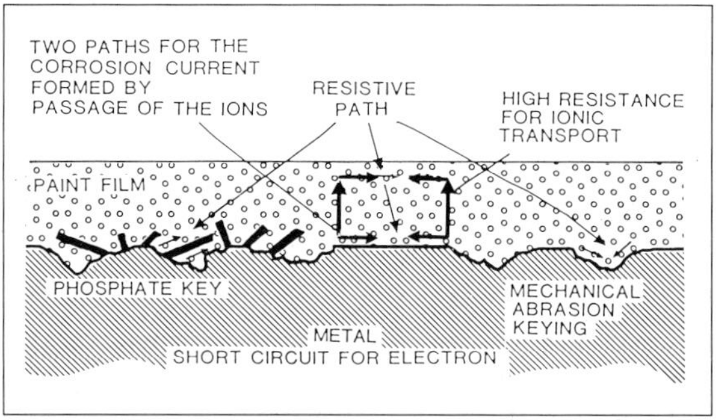

FIGURE 8. A paint film stops corrosion by reducing the ionic current flowing between anode and cthode, (figure 4) (7.5), by providing an adherent film which increases the electrolyte resistance between the anodic and cathodic sites. This figure depicts the resistive path for: a paint film; a phosphate film, a mechanically abraded surface, a flat untreated surface. The resistive path for the ionic current (figure 4) through the paint film is very high, and so the corrosion current will be very low. If you are unfamiliar with electrical resistances, you can consider that the paint 'insulates' the metal from corrosion to a certain degree, but see (2.4, 2.6, 1.10).

is often recommended to achieve this adherence. In practice, this roughening and keying is achieved on the vehicle production line with a chemical reaction (employing *phosphoric acid (4.2)), which requires water rinsing. The drainage holes, (plates 4, and 11), provide the water exit path for this industrial process, (4.2).

* ***Phosphoric acid*** is used in rust treatments, (which can be purchased from DIY motor factors).

2.2 PAINT FILMS

The main functions of a paint film is to:

a) be adherent to the metal because both anodic and cathodic reactions, (figure 4), must take place on the metal surface. These reactions cannot take place at the paint/air interface;
b) be coherent within itself with no voids. This is a requirement so that the de-icing salts don't get to the metal/paint interface;
c) eliminate the oxygen in water 'charge' (1.8) by providing equal replenishment of oxygen over the whole (wetted) metal surface of the car;
d) exclude the aggressive de-icing salts and their associated environment;
e) provide lustre for decoration;
f) be of uniform thickness;
g) hide the roughness associated with the keying phosphates.

2.3 PROBLEMS OF PAINT FILMS IN CORROSION

Unfortunately not all the surfaces on the car body can obtain an even protective coating so as to provide equal oxygen replenishment rates at the steel surface (figure, 3). Crevices, metal overlaps, spot welded seams (figure 2) and interwelds (3.6) are such areas.

2.4 PASSAGE OF WATER THROUGH A PAINT

What most people will find surprising is that a paint film will allow as much pure water to pass through the film, as an uncoated steel surface needs to corrode. Therefore the protection mechanism of a paint film is **not** the physical exclusion of water from the steel surface but by reducing the ionic current (figure 4, (1.4)) which has to flow within the thin 'liquid' which exists at the metal and paint interface.

2.5 CORROSION PROTECTION BY PAINT

Paint films are impermeable, over short periods of time, to dissolved salts like the road de-icing salts. The protection offered is proportional to the paint film thickness. On the external vehicle there are several coats of paint, but only one coat on the internal surfaces of 'enclosed out of sight' panels. The coating thickness on these panels is very small as will be explained in chapter 4. It has been the practice to increase the durability of thin coatings by the addition of waxes (6.2).

2.6 PAINT ADHERENCE TO THE SUBSTRATE

The chemical by-products of the corrosion process destroys the adherence of the paint film to the metal substrate producing ***scab corrosion*** (figure 9 and 10). NB. These chemical by-products are all produced at the metal surface between the metal/paint interface and this causes the paint film to be blown up by the imbibed water, ie water which physically passes from the paint/air surface to the metal/paint interface as shown in figures 9 and 10. Many painters will interpret this as the 'metal sweating'.

2.7 NON ADHERENT PAINT FILMS

Non adherent paint films should be removed because they can trap water as described in the previous section. As soon as the paint film loses adherence to the metal, it can not stop the lateral spread of corrosion, nor can it provide equal oxygen replenishment to the whole 'wetted' metal surface. The profile of the non adherent paint film is such that it causes an unequal concentration of oxygen over the wetted surface (figures 6b, and 9), and therefore provides the driving force for the corrosion of the substrate metal at the centre of the paint blister. This difference in oxygen availability becomes more marked (figure 10) as the corrosion front progresses into the metal, which in turn promotes more corrosion at the

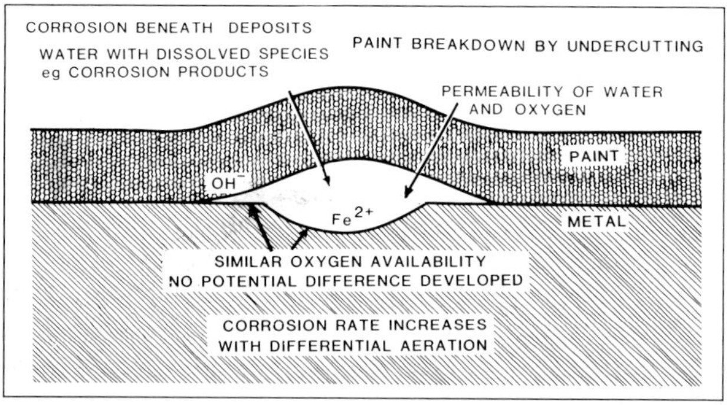

FIGURE 9. A paint film also provides similar oxygen replenishment to the whole surface and therefore the whole surface remains as a latent battery. However dirt particles, which remain in contact with the paint film, (or non adherent paint films), cause an unequal concentration of oxygen and therefore provide the driving force for the corrosion of the substrate metal. Corrosion proceeds as the alkali strips the paint film causing a loss of adherence to the metal. The non adherent paint film becomes full of water. To differentiate between this defect and metal perforation, (both outwardly look like 'L' on plate 1), refer to figure 25. (Proceed to figure 10).

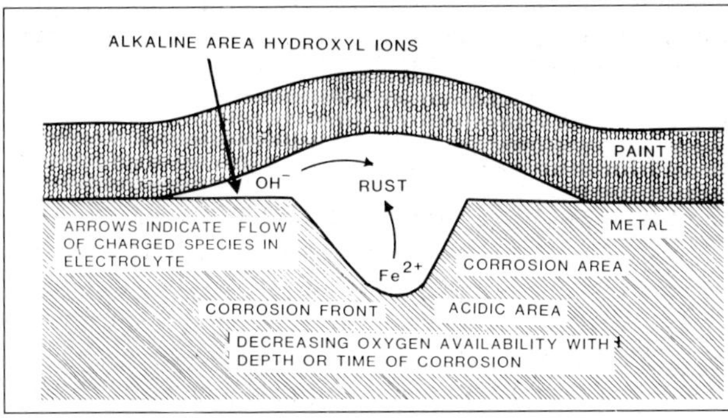

FIGURE 10. The corrosion rate increases in proportion to the difference in the oxygen availability between the two sites (anode and cathode). As the corrosion front progresses into the metal it becomes further starved of oxygen, therefore ***increasing*** the current from the short circuited battery. The corrosion front (or pit) develops an acidic environment (1.10) which accelerates the corrosion process.

centre. This corrosion process develops a pit. Once localised pitting has started, it is very difficult to stop as subsequent paint films will bridge the pit. This explains the requirement of the corrosion warranty that the owner should rectify all surface nucleated corrosion. ***Failure to undertake this remedial, protective maintenance will invalidate the 6 year warranty*** (8.6).

2.8 SUMMARY. Environmental durability of paint films.

No protective, decorative paint film (even if it cost as much as gold), will withstand the long term contact with:

a) internally nucleated corrosion, ie rust through penetration;

b) corrosion which is due to stone chips flung up from the road wheels (figure 14). Stone chips would tend to leave areas devoid of paint due to the impact of the stone. Most paint damage, (which is wrongly ascribed to as ***stone chipping***), is due the erosion of the paint film by small water droplets (figure 16), which impact the painted surface as a result of bad design, (see 8.6);

c) running water (plate 1) either as rain or in the road spray, which perpetually makes contact with the same regions of the paint film;

d) the alkali produced by the cathodic reaction of a corrosion reaction which can be nucleated at a crevice (1.6), such as an interweld (figure 3), acts as a paint stripper to the external paint system. Large areas of paint can be stripped by this means;

e) stationary road de-icing salts in contact with the paint for long periods of time will damage the paint-work and lead to 'rust through'. This usually occurs on internal panels which are 'out of sight of the owner', and form the basis of the manufacturer's 6 year corrosion warranty;

f) mud compacts, because mud pockets hold a reservoir of water;

g) plastic adjacent to the surface of painted steel.

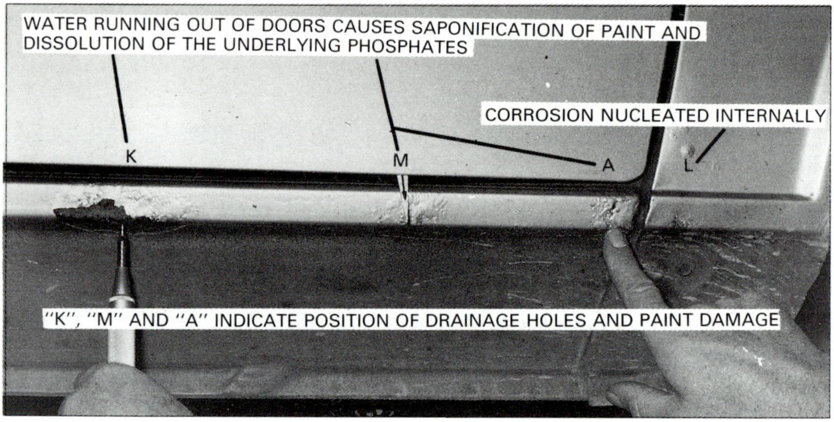

Plate 1. View of lower door, sill and front wing, showing two defects, neither of them can be attributable to stone chipping as was emphatically stated by the manufacturer's agents. (**1**) The paint blister 'L' shows the result of early rust through of the metal panel forming the front wing (detection figure 25). Note the waviness associated with crevice corrosion along the vertical clinch fold (figure 13). (**2**) The water entering the front door through incorrectly designed door glazing rubbers (plate 16), causes paint problems by three mechanisms as it runs out. These mechanisms are: (**a**) by dissolving the underlying phosphates on the sill as the water runs out of the door drainage holes at 'K', 'M', and 'A'. These letters coincide directly with the door drainage holes; (**b**) by corrosion. This waterfall forms ***perpetual*** cathodic and anodic sites; (**c**) the alkali produced (1.10) by the cathodic reaction of this corrosion process, then acts as a paint stripper to the external paint system. The non adherent paint then swells up by imbibing water (2.6), (figures 9 and 10).

Any one of these conditions is enough to break down the protective paint film(s). Naturally, after discussing plate 1, you will want to know how the paint on the roof withstands rain water. The secret to understanding this point is in the ***perpetual*** nature of certain droplets which cause corrosion compared to the ***random*** nature of other droplets like rain. The random droplets have the ability to change anodic sites to cathodic sites and vice versa. Yesterday's anodic site will come to-day's cathodic site. Therefore the corrosion battery does not fully develop, (1.10). Another point is that the rain droplets, although they are stationary, they rapidly dry. This has to be contrasted with the slower drying which accompanies dirt deposits at their base.

CHAPTER 3

Body Build. Safety in Accidents

3.0 BODY BUILD. The stressed body as a 'chassis'.

The modern mass produced car is built from steel which is supplied as sheet. From these sheets, panels are contoured and formed to shape and then spot welded together to form the load carrying structure (figure 2). Every contour increases the rigidity of the sheet metal. The strongest section which can be made is a ***box member***. This can be demonstrated by forming a piece of paper, which when loaded as shown in figure 11a, will not carry any weight. When the same mass of paper is formed into a box and glued, (figure 11b), it will carry a substantial load provided the glue line remains intact, and the rectangular shape is maintained. A greater load can be carried if the box member is cross braced (figure 11c) so that the rectangular shape is guaranteed thereby making it a compartmentalised enclosed box.

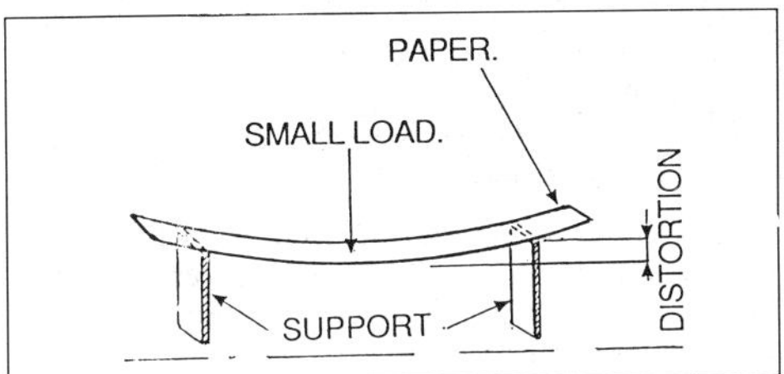

Figure 11a

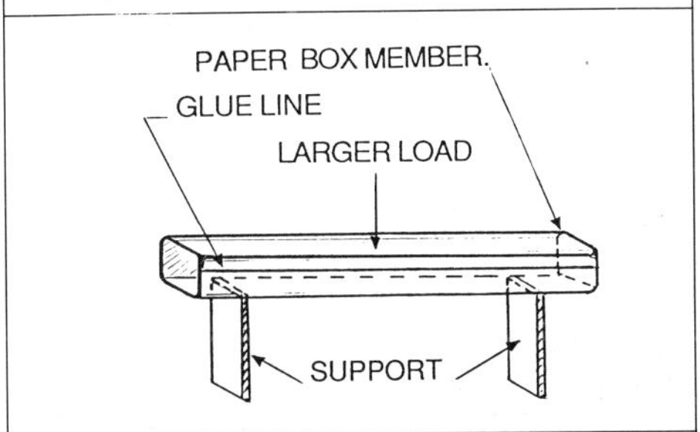

Figure 11b

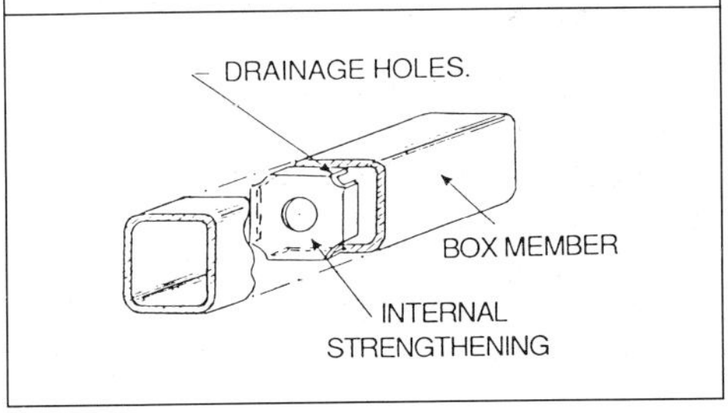

Figure 11c

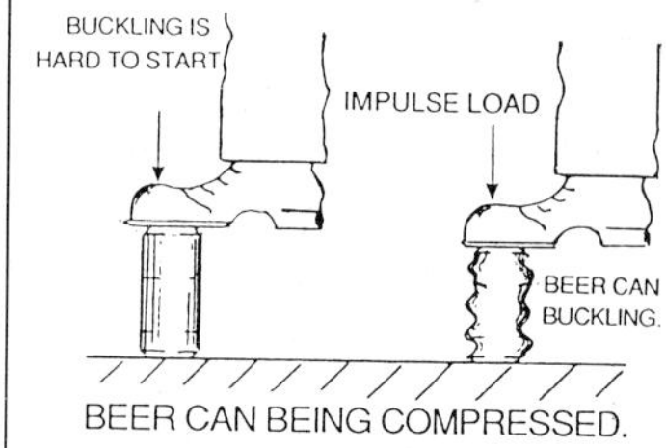

Figure 11e

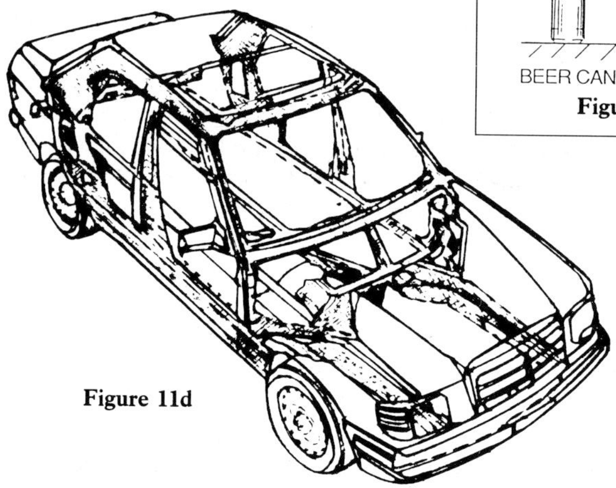

Figure 11d

FIGURE 11d and e. The principles shown which the industry uses to make the modern car from sheet steel (figure 2). The energy forming structure is shown shaded in figure 11d. Figure 11 drawn by Maurice Neal and acknowledged with thanks.

For corrosion protection, all compartments and surfaces of these box sections, and strengthening regions, must be painted and the glue line, (equivalent to the line of spot welds), must remain intact. These principles are used to increase the rigidity and stiffness of the car's structure. The ***box members*** of the body shell now performs the functions of the ***chassis*** and carries all the stresses, (shaded areas in figure 11d). The integrity of this boxed safety cell is very important when it comes to surviving in an accident (3.5). This method of forming the stress carrying body is called the unitary or monocoque construction and relies upon continuity of stress paths throughout the *whole* body.

3.1 LOCALISED REINFORCEMENT

Where the stresses are ***highest***, at fixing points to the engine and wheels, there is considerable local strengthening. This often creates large metal overlaps, (crevices), which hold a large amount of water by capillary wicking (figure 21a). They dry slowly, and therefore give rise to potential corrosion sites.

3.2 COSMETIC PANELS and LOAD BEARING PANELS

Certain panels which do not transmit any stresses are called cosmetic panels; these are the ***wings, bonnet, doors and boot***. The condition of cosmetic panels is excluded from the MOT examination because they are not regarded as stressed safety members. These cosmetic panels ***do*** come under the manufacturers metal perforation corrosion warranty. Not all stress carrying box members are examined at the MOT examination. An example of this anomaly is the 'A' pillar into which is bonded the windscreen glazing (figure 11d). This method of construction increases the torsional rigidity of the vehicle body and produces a slimmer windscreen 'A' pillar allowing for better forward vision. The 'A' pillar/windscreen area is also excluded from the after market corrosion warranties, but is included in the manufacturers warranty.

3.3 SAFETY LEGISLATION

Further additional requirements are made of the vehicle body. Government legislation requires all passengers to be given a chance of surviving an impact with another car or structure. New vehicles undergo extensive tests to ensure that box members surrounding the passenger cell are ***almost non deformable***. The occupants are anchored to this cell by virtue of their seat belts, and the front and rear structures must absorb some of the kinetic energy of the car during accident time. This is achieved with the use of steel and designed in 'crumple zones'. Consider again our paper box and let us compare this with an empty steel beer can. We know that if we want to squeeze the beer can flat, we have to stamp very hard on the can supplying a downward impulse onto the steel can (figure 11e). If this downward impulse is small, (ie its kinetic energy is small), the beer can will absorb the impulse and remain only partially squeezed. It achieves this because the steel gets stronger, as it is deformed and concertinaed. It is for similar reasons that a paper clip can never be straightened due to ***work hardening***. If the empty beer can were partially cut along the axis, there would be little or no energy absorbed in its' compaction. The deformation in this state will correspond to the corrosion within the crevice between spot welds of a vehicle's box member. Paper, cardboard, or glass fibre on the other hand, readily collapse into a flat pack and absorb little energy. These materials are forbidden as repair materials for *stressed members, (*called ***prescribed areas*** in MOT jargon, (9.1)).

3.4 SAFETY DURING ACCIDENT TIME

The function of a vehicle designer is to produce a rigid structure, which in the event of an accident, collapses progressively with ***accident time*** absorbing the ***energy of the accident***. Indeed, some designs of vehicle bodies have incorporated in their structures artificially induced "crumple zones" which achieve this progressive collapse. An important question is: ***At what age does the vehicle body lose this progressive collapse with accident time as a result of corrosion?.***

3.5 IS THE VEHICLE SAFE AT ANY AGE?

Figure 12 shows that there may be a correlation between age of vehicle with the ability to protect the occupants. In this graph, the number of occupants who become 'accident victims' when involved in an aging vehicle are correlated with age of vehicle. I believe this correlation is due to the loss of this ***progressive collapse*** as a result of corrosion. The two graphs are the result of the Swedish MOT test station reports (***Graph line A***) and the Highway Safety Research Centre (***Graph line B***). ***Graph line A*** relates the age of vehicle to the percentage of remarks made about corrosion in the ***force transmitting members***. As one would expect, these increase with age of vehicle. ***Graph line B*** superimposes upon graph A with a common age axis, the ***overall accident rate per million miles*** and suggests that the occupants in an older car are more likely to be injured in any accident as a result of this premature corrosion (Figure 12). It is important to emphasise that corrosion is not the cause of the accident, but

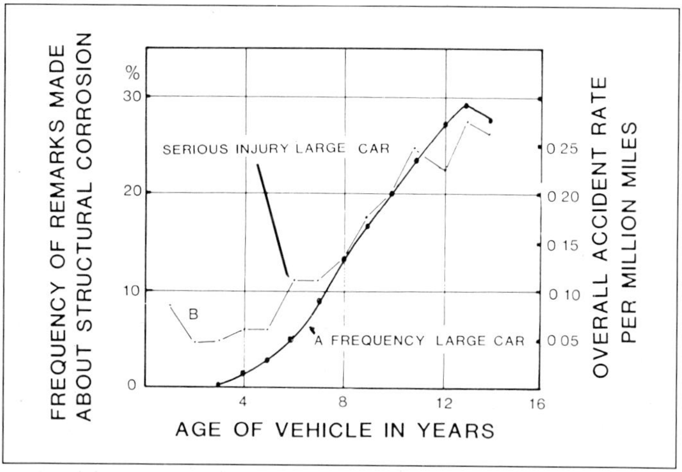

FIGURE 12. Relative frequency of remarks concerning frame and structural corrosion of the motor vehicle superimposed upon the crash rates per million miles with regard to the age of vehicle. The curve is discussed in the text. The sources of data used are: ***Curve A*** Svenson G E Swedish Corrosion Institute (private Communication) also published in "Cars go Rusty" AB Svensk Bilprovning. ***Curve B*** Dutt A M & Reinfurt D A "Accident Involvement and crash injury rates by make, model, and year of car". Highway Safety Research Centre University of North Carolina DOT-HS-4-00897. Private communication Reinfurt D A.

rather the reason why the occupants were admitted to hospital. They were admitted due to the injuries received as a result of the non progressive energy absorption with accident time. Several explanations are available for this trend shown in figure 12. These are:

1) the automotive designer is achieving his goal in that the younger vehicles are more crash-worthy;
2) the older vehicle being cheaper, is run by young people (18-25 years old), who are more accident prone than the older person, and who can not afford to run a new car;
3) that corrosion is affecting the ability of the steel to absorb energy during the 'accident time' as a result of interweld corrosion (figure 13, (3.7)). The energy absorbing front structure easily collapses providing little retardation. There is rapid deceleration when the engine comes into contact with the bulk head causing very high 'g' forces which impart some further injury to the occupants.

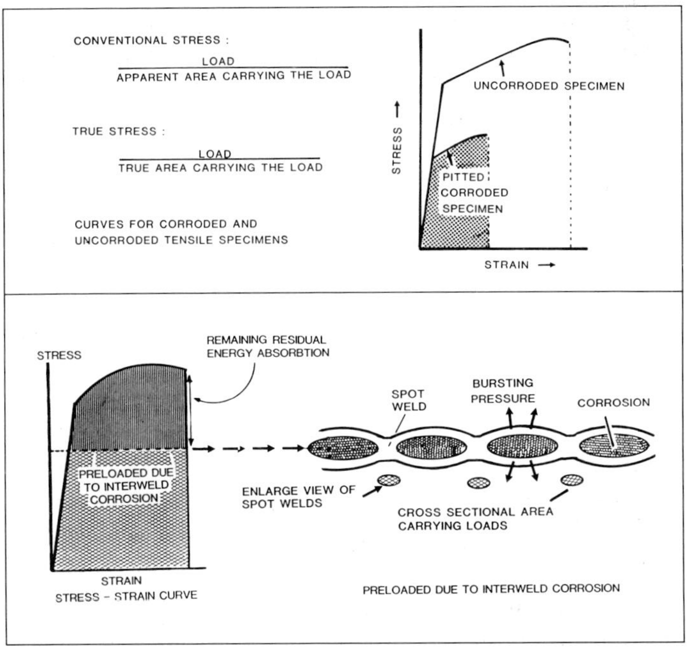

FIGURE 13. (**a**) The shaded area beneath the stress strain curve is proportional to the energy absorbed by the sample. (**b**) An easier way of considering the same concepts is that corrosion within the interwelds will pre-load the interweld by a process of 'oxide jacking' caused by the increase in volume occupied by the rust. The energy of absorption is now much less because of this pre-loading which shifts the origin vertically upwards by the pre-loading forces.

3.6 THE INTERWELD

The line of spot welds and the associated crevices, (called the interweld), is an important part of the vehicle's load bearing structure. The integrity of the interweld is also important because it allows the steel box member to absorb kinetic energy thereby reducing the deceleration forces experienced by the occupants during the short period of the accident. A continuously welded crevice would be too rigid and too expensive to make. If your 6 year corrosion warranty excludes seam or interweld corrosion, then the corrosion warranty is not worth the paper on which it is written, (see repair of interwelds 10.19). You are better off not buying the vehicle as it will turn out to be a lemon.

3.7 RUPTURE OF SPOT WELDS BY CORROSION

Corrosion effects the ability of a steel specimen to withstand high stresses (figure 13a). This is particularly true when the corrosion takes place in the interwelds. The corrosion product formed, the oxide ***rust***, occupies a much larger volume than was occupied by the metal which has gone to form the rust. This expansion will ***pre-load*** the interweld by a process of 'oxide jacking' (figure 13b). Interweld corrosion will be shown clearly by the 'wavy' or bulging interweld seam. The energy of absorption (measured by the shaded area under the stress-strain curve) of the components affected by interweld corrosion (our glue line) is now much less because of this pre-loading which shifts the origin vertically upwards.

GLOSSARY (continued from page iv)

Posts or Pillars. The vertical pillars are given prefixes **A, B and C**. **A** for the windscreen pillar, **B** pillar for the pillar upon which the rear passenger doors are hung and the **C** pillar for the door closure over the rear wheel arch.

pH is a measure of the acidity (or alkalinity) of a liquid. Pure water has a pH of 7 and an acid has a pH of less than 7. Alkalis have a pH of greater than 7.

Throw (of electrolysis, or electroplating) is the extent to which the current may reach the opposite electrode and deposit the metal plating or paint.
Saponification. Plate 1, (2.8) Saponification is a term used to express the degradation of a paint by alkalis. The paint is turned into a 'soap', hence the name.

Thixotropic. A property of gelatinous liquids which do not drip, but become more like a liquid when shaken, stirred, or sheared (on spraying), eg non drip household paints are thixotropic.

Terminology. American, Australian and European terminology for parts of a car are different. In this book, English terminology has been used. In the following, the English usage in bold, European and American usage in normal type eg;

Boot-Trunk. **Bonnet**-Hood. **Sills**-Rockers.

CHAPTER 4

Mass Production Painting Process

4.0 MASS PRODUCTION PAINTING SYSTEMS

In this chapter, I want to lay the basis of the predictive element of corrosion so that a unified theory can be applied to all monocoque vehicles (including commercial vehicles) from 1950 to the present day. This requires a knowledge of the painting process, (4.10).

4.1 HISTORY OF CAR PAINTING (for restoration enthusiasts)

The history of mass produced car painting has evolved as follows:

Prewar vehicles were built on a chassis and were spray primed. All areas were accessible and crevices were usually filled with beading. With the introduction, in the 1950's, of the all welded chassis described in chapter 3, solvent based dip primers, followed by water based dip primers were used. The vehicle bodies were immersed into the paint bath and came out painted. If you painted an article in this way you would expect runs and sags in the paint. Runs and sags were a characteristic of both the solvent and water based dip primers. These had to be rubbed down by hand before the final top coats could be applied. Principle (4.2) applies but large drying holes to remove the solvents took the place of the holes described in plate 5, (4.5). Solvent washing (•) of the internal box members was a problem which gave rise to thin coats of paint on these internal members.

From 1960 the water based anodic (•) (1.5) electro-coats were introduced which gave a greater production rate, and eliminated these defects (runs and sags), and therefore reduced the amount of hand preparation required. In 1977 the cathodic(•) electro-coats were used. The majority of vehicles are painted by these electro-coating techniques.

4.2 THE BODY PROCESS DRAINAGE HOLES

The public in general have the wrong conception of vehicle drainage holes. Most readers would agree that their concept of drainage holes arises from the mass media which gives the drainage holes as the means by which road spray and water is drained should it get into the vehicle box members, (the manufacturers reinforce this concept by advising that the drainage holes should not become blocked, see plate 4). This is not the truth as they are solely to let out the manufacturing liquids; degreasers and their rinses; pretreatment phosphating solutions (2.1) and their rinses; paints and their rinses. In this section, I want to give an insight into the vehicle painting system and describe the function of the drainage holes where painting is concerned. The vehicle body is spot welded together, (figure 2), and painted in large paint baths by total immersion (plate 2). The car body must;

a) sink in the paint bath, and

b) satisfy certain conditions when leaving the bath. The paint must run out uniformly (to reduce 'carry-over') and yet still maintain a film on the surface. When a milk bottle is filled with water and inverted, it provides a very turbulent flow when raised out of the water bath. If this same turbulence were allowed to happen when the car left the bath, most of the paint film would be removed around the orifices of each hole.

For both conditions (a) and (b) to be achieved, trapped paint/or air must escape slowly and freely so that the flow of the emerging paint is unrestricted. Therefore ***all*** box members have ***air admittance and exit holes*** (plates 3 and 18) and ***paint*** drainage holes (plates 4, and 11). These

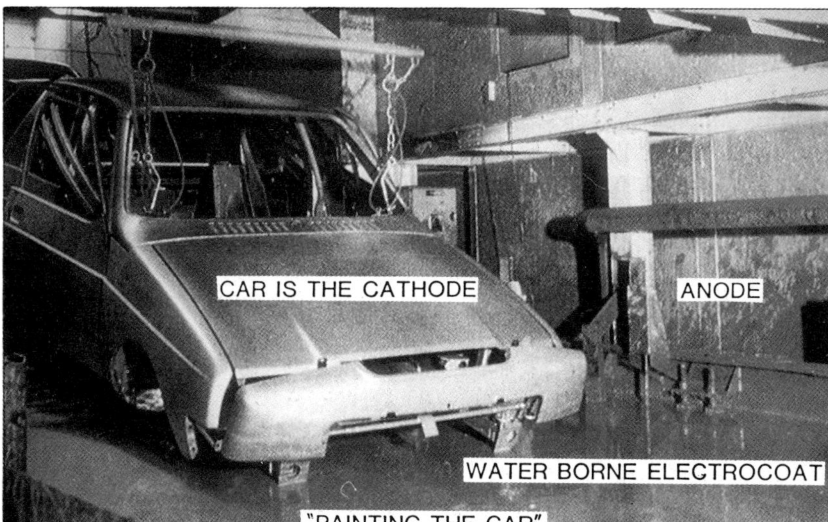

Plate 2. Shows a car entering a cathodic electro-coat paint bath. The car is suspended from an overhead track, which gradually lowers the vehicle into the bath at the angle shown here. In order for the car body to sink in the electro-coating bath, air must not be trapped in hollow box sections. Conversely, on emerging (front first and at the same speed and angle), air must be admitted to the hollow box sections in order for the paint to run out slowly and uniformly. Hence there are air admittance and air exit holes at the rear and front section of the horizontal box members (see Plates 3 and 18). Other factors which determine whether a vehicle body floats in the paint bath are: body design, body sling design, conveyor speed and angle of immersion. Note. The same principles apply to the ***all welded monocoque*** body (3.0) when painted by immersion techniques, (1950-present day). Acknowledgement: Plate contributed by Citroen Cars Publications and acknowledged with thanks.

holes are the manufacturer's ***process*** holes, and to avoid confusion the term ***drainage*** hole will only be used to describe those holes required for drainage of the manufacturer's process liquids (plates 4, 11, 12). Bolted on components are also painted by immersion techniques; see also (5.11, 5.14, 5.20, 5.21, figure 23).

4.3 PAINTING (by ELECTROCOATING 1.5)

The electro-coats were introduced to reduce the amount of hand preparation required to obtain a level run and sag free primed surface. The property of this electro-coat paint film is unlike the general public's understanding of a paint film. First of all, immersion of the car body in the water borne paint, does not produce an adherent and coherent paint film on the surface unless it is chemically deposited by an electric current. While the car is immersed in the paint, which is a water borne paint, to reduce the risk of fire, it is subjected to an electric current, in rather the same manner as the plating of metals in electroplating. The paint is electrically deposited as a 'wet powder', which is subsequently heated to liquefy the paint to provide a continuous film as we know a paint film.

4.4 THICKNESS OF THE ELECTROCOAT

The film deposited by electrocoating is a non conductor of electricity and so the ***coating acts as its own thickness regulator***. The thickness of the paint film varies on the car body depending on how far away the panels in question are from the opposite (or counter) electrode, called anodes (plate 2) for cataphoresis (1.5).

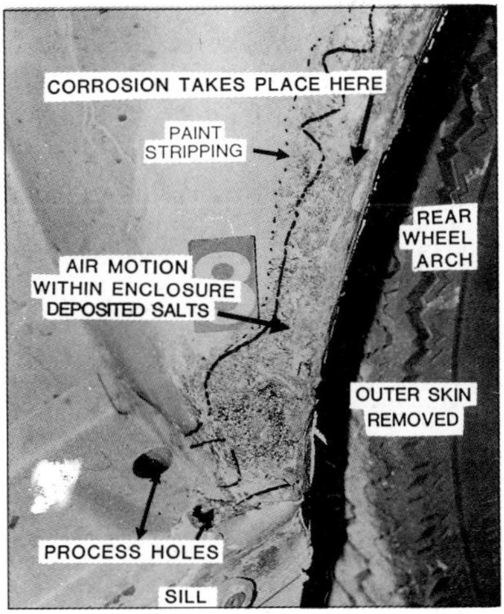

Plate 3. The outer panel of this rear wheel arch has been removed to show the process holes which perform the function of allowing air to escape from the box section beneath, see also plate 18. The air movement resulting from the passenger air extraction system (5.12 to 5.13) is clearly shown by the corrosion beneath the deposits over the rear wheel arch. This air has been drawn from the front wheel track, through the sill/rocker drainage holes (plates 11 and 12) into the sills/rockers and then through the air exit hole shown in this plate. Throughout the passage of this front wheel spray, the low momentum spray has deposited the road de-icing salts which adhesively bonds the dirt particles onto the vertical, (internal, and out of sight), panels when they dry. **DIY rectification** see (10.7.3).

Plate 4. Shows two defects: (1) An example of a paint drainage hole placed directly over the front* wheel track which ceases to act as a drainage system as it collects road spray and becomes full of road dirt. Refer to figure 20 for a descriptive analysis. **DIY rectification:** This drainage hole should be fitted with a one way valve (Figure 17 method B). (2) The door bottom is proud of the sill and so the drainage holes at the bottom of the door collect the vehicle's own front wheel spray (figure 21(B)). This results in localised corrosion around the drainage holes (plate 13). **DIY rectification:** fit mudflaps. *Same principle applies to the drainage holes behind the rear wheels.

4.5 ELECTROCOAT THROW HOLES

Like the metal plating baths in electroplating (1.5), the paint remains in solution if no current flows. The terms used in electrocoating are similar to the terms used in electroplating. Hence the term ***'throw'*** or ***'throwing power'*** is used to describe how well the paint or metal is deposited at large distances from the counter electrode. The 'throwing power' of the electrocoat process is also used to describe how well the paint is deposited upon the internal surfaces of box members and interweld crevices. Therefore, for the paint to be deposited on the internal surfaces of box/enclosed members to any appreciable degree for protection, electrocoat 'throw holes' are required, (denoted by (A) in plate 5). Their function is to increase the current flow within the enclosed 'box' members and allow the electric current to deposit and 'fix' the paint upon the internal surfaces. Hence the passenger car compartment is linked to the external pressure eddies beneath the car by these process holes (electrocoat throw holes (plate 5), paint drainage holes (plate 4), and air displacement holes (plate 3)). For post war cars, before the advent of electrocoating, drying or ventilation holes were required to prevent the washing away of the primer from the vertical surfaces by the condensing solvents.

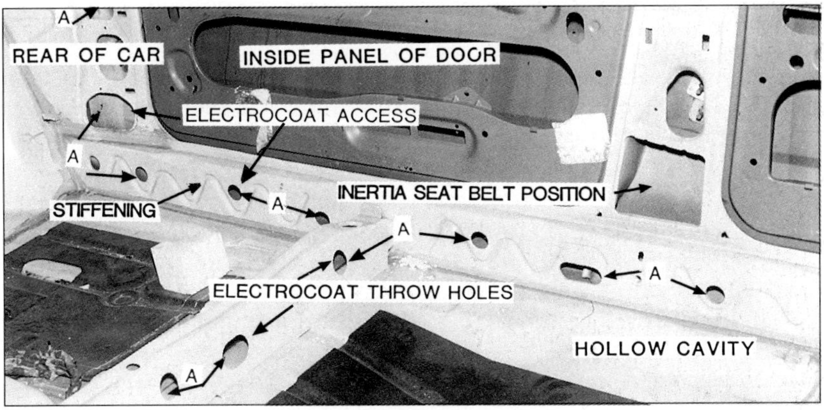

Plate 5. A car body with the internal decorative panels removed, showing the holes which are required to achieve a better electrocoat paint deposit. These electro-coat throw process holes ('A'), link the passenger compartment to the dense water spray over the front wheel track, (via the drainage holes in the sills/rockers, plates 4, 11, and 12). These holes (A) can also be used for DIY wax injection and therefore eliminates the need for drilling the sills and rockers.

4.6 ADVANTAGES OF ELECTROCOATING

Electrocoating is the most foolproof way of painting an intricate shape (made from a metallic conductor of electricity), as there is no 'shadowing' as there is with a conventional spray gun. In the deposited form, the paint is not affected by runs and sags which you would get if you tried to do exactly the same immersion test with domestic paints. Furthermore the electrically deposited 'paint' can also be washed with water rinses without being either washed away or diluted. Imagine how easy domestic painting would be with the same properties: own thickness regulator, no runs, no sags, no fire hazard, does not dissolve again in its solvent, but can be rubbed off with a dry cloth until liquefied in a subsequent baking process.

4.7 OUT OF SIGHT PAINT COATS

The electrocoat paint film is all the 'painting' which the internal and underbody panels are given. Underbody panels are also covered with a chip resistant *PVC plastisol coating adjacent to the front wheel tracks.

* PVC is short for poly vinyl chloride. The plastisol is a very thick fluid which is sprayed onto the electrocoat and performs the same function of underseals. It is very thick and does not creep into crevices, (figure 22).

The outside panels are given several more coats and these are the ones which are heavily advertised in the press. When the painting of the vehicle is mentioned in advertisements and other brochure material, the underbody coats are not fully discussed and only the outside (thicker) coats are discussed in any depth. It is left to the reader to believe that the same layers are given to the underbody.

4.8 THE PAINT THICKNESS IN INTERWELDS

The interweld region is denuded of electro-paint (figure 3) due to the very small electrocoating current which is available to deposit the paint. A similar effect would be observed in the electroplating of metals in narrow crevices where there has to be a current flowing for the paint (or metal) to be deposited. The important principle is ***no current*** flowing means ***no paint.***

4.9 HOLIDAYS IN THE COATING

An analogy of the paint throwing process can be a stream of passengers entering an underground train system. As the passengers come down the steps and escalators, they must pass through a narrow door to get onto the platform. The narrow doorway is analogous to the small electrocoat throw holes (A in Plate 5). As the passengers pass through this narrow door way they can spread out along the platform facing the line (equivalent to the paint being deposited). Human nature being what it is, they will stand with a gap between each passenger. In painting technology, these gaps are called ***'holidays'***. The passengers will stand preferentially nearer the doorway. This is exactly how the paint is deposited, more paint is deposited nearer the electrocoat (throw) holes than further away from these holes. There will be more queuing and double and triple queues nearer the doorway, when perhaps there are gaps (holidays) within the passengers further down the platform. As more passengers enter the platform, these 'gaps' will be filled, and slowly the passengers will all stand against the line. Further passengers will then start to queue behind the first arrivals. This is analogous to the paint film ***thickening up***. Ideally the paint film does not thicken up until all the holidays are covered. This does not happen in practice as shown in figure 3.

4.10 SUMMARY

All mass produced welded bodies painted by immersion techniques require the vehicle body to contain ***process holes,*** (air admittance holes, air exit holes and paint drainage holes) and either electro-coat throw holes or solvent drying holes. Hence the passenger car compartment is linked to the external pressure eddies beneath the car by these process holes, and therefore the principles given in chapter 5 applies to all monocoque vehicles (including commercial vehicles) from 1950 to the present day.

CHAPTER 5

Mechanics of Spray Ingress

5.0 MOTORWAY SPEEDS

The average motorway driving speed is increasing every year. In one second the car travels 88 feet at 60 mph (figure 20). The roads are impervious to water and this means that the slightest rainfall provides an environment of 'recycled rain', which in the winter, is essentially recycled sea water spray (figure 1).

5.1 ORIGINS OF ROAD SPRAY

The following points apply to both front and rear wheels. Figure 14a shows the origins of vehicle spray which is developed by the following processes:

1) The 'bow wave' produced when the tyre tries to push a finite volume of water ahead of the tyre. This spray is responsible for side impacts on the car body (5.10).

2a) Spray is produced upon the separation of the tyre from the road (5.4).

2b) Spray is also formed as the water is thrown out of the tyre tread over the whole circumference of revolution. These droplets impact the vehicle body at the vehicle speed (5.2).

3) Droplet-body impact. The water droplets which are just lifted off the road (and have zero velocity with respect to the road) produce a very fine spray (figure 15) on impacting any part of the car body which is perpendicular to the direction of motion or impacting wheel spray. The importance of this is for jacking points (plate 17) and other protuberances (plates 10 and 18, figure 23) beneath the vehicle. The change of energy from kinetic energy (of one drop) to surface free energy of a number of micro-droplets of small momentum causes a fine cloud. These micro-droplets follow the pressure eddies around the vehicle body.

   Note certain members (eg plate 18), have the same reinforcing action if they are placed underneath the vehicle floor pan as they do when mounted within the passenger compartment plate 5. The vehicle design dates are later for plate 18 than for plate 5, which shows that designs can get worse with time.

4) Droplet-droplet impact with the spray from both front and rear wheels. These collisions ahead of the rear wheel produce spray. Some spray is also produced by the spray from the front wheel colliding with the rear wheel at the point (4) in figure 14a.

5) Water spray thrown ***in front*** of the vehicle at ***twice*** the vehicle speed, (5.2), collides with the underbody. The relative speed at collision is the vehicle speed.
6) The preceding vehicle causes a turbulent spray through which other vehicles must travel. This applies to overtaking and overtaken vehicles alike.

5.2 SPRAY FROM ROTATING WHEELS.

The ***top*** of the wheel throws spray off in a forwards direction at ***twice the vehicle speed*** relative to the road. This principle is not considered by many designers and leads to the provision of ***partial*** wing liners (figure 14b) and the design fault shown in plate 6. The easiest way to show that a rotating and travelling wheel throws spray off the top at twice the vehicle speed is to consider a bicycle wheel held just touching the surface of a water bath. The axle is stationary, but the wheel will shoot spray

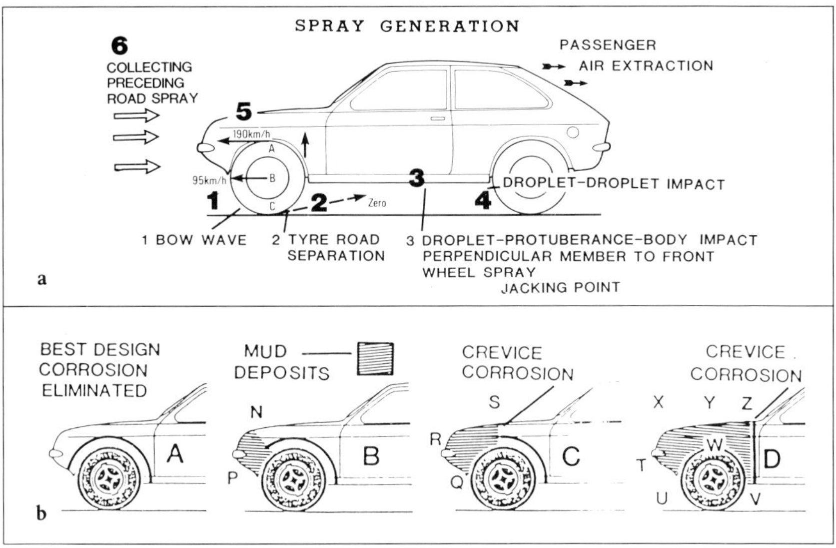

FIGURE 14. (**a**) Shows the 6 methods of spray generation, and the analysis of the tyre throwing road dirt forward of the vehicle at twice the vehicle speed. (**b**) Shows four possible wing liners or stone guards made by different vehicle manufacturers and fixed beneath the wings of the motor car. Only the type depicted in (A) is satisfactory from an anti-corrosion design point of view. This results from the water thrown forward of the wheel by the rotating wheel (see also plate 6). The reader should look at the current models with this in mind. Design considerations should incorporate adequate fixing points around the wing in order to eliminate the possibility of crevice corrosion between the metal and plastic junction. There should be no possibility of the fine spray by-passing the edges of the guard and resting on the inner wing panel, from where it cannot be washed away.

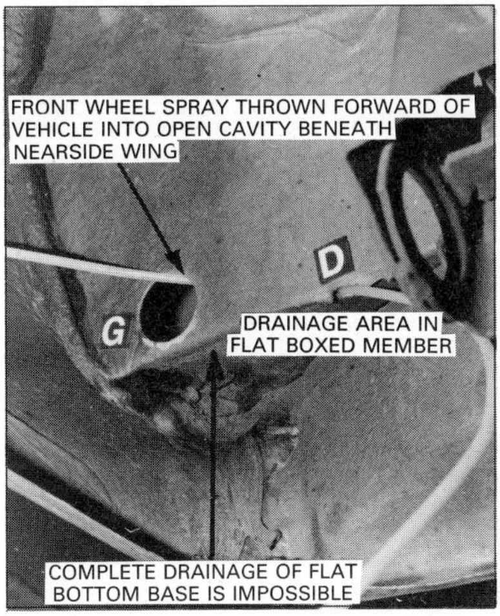

Plate 6. A view of the near side front* wheel arch from beneath the car. The photograph is taken looking towards the front of the car. This box member forms a structural energy absorbing member in accidents. The rotating front wheel, (removed for photography), throws spray from the top of the wheel at twice the vehicle speed (figure 14a). This will be collected by the hole G, (which only exists on the near side front wheel arch). The small drainage hole, 'D', will become blocked due to the poor drainage of the road dirt which is deposited internally upon a horizontal surface. This cavity box member, (behind the panel GD), has a panel which forms a perpendicular air dam to the preceding vehicle spray. This panel can be seen from the front of the car, and therefore there will always be disagreements between the warranty agents and yourself as to whether the paint blemish (like 'L' on plate 1), is the first signs of internally nucleated metal perforation corrosion, or is it non adherent paint (figure 9) resulting from stone chips. Most owners will settle for stone chips. You should test for metal perforation (figure 25 for test method and figure 24 for the correct orientation of the pin where double skins, interwelds etc are concerned). These large holes [G] were described by agents acting on behalf of the manufacturer, as 'ventilation holes for the structure'. **DIY Rectification**: These holes should be fitted with a one way valve (figure 17 method C). *Same principle applies to the open or wide crevices and seams in front of the rear wheels (eg sill closures).

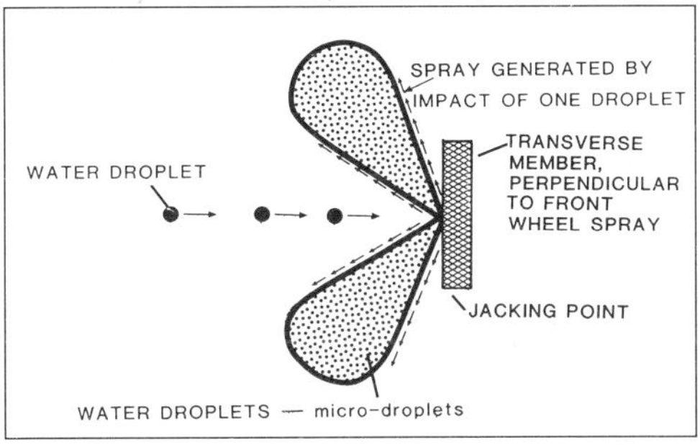

FIGURE 15. A body member (eg plates 10, 17, and 18) at right angles to the vehicle's direction of motion will 'collide' with the front wheel road spray (figure 14a mechanisms 2 and 3). This is a schematic diagram showing the spray generated when one droplet is turned into a cloud of micro-droplets. This mechanism of spray generation is important for the water ingress into sills/rockers through the holes associated with the jacking points. This is important and is **not** acknowledged by the motor vehicle manufacturers at the design stage, (eg open holes R and U plate 18). Other defects which could be illustrated are: protuberances associated outward flared wheel arches; doors not aligned with the 'C' post; internal stone guards to rear wheel arches protruding below the sills and thus collect the front wheel spray; and factory fitted mudflaps positioned adjacent to drainage holes.

backwards. If the wheel must travel forwards, then the point of rotation becomes momentarily the interface between the tyre and road (point C in figure 14a). This means that the axle (B figure 14a) travels at 60 mph, whilst the top of the wheel (A) is travelling at twice the vehicle speed, because the centre of rotation is now the point of contact with the road and the equivalent radii of rotation are the wheel radius (BC for the axle) and twice the wheel radius (AC for the top of the wheel). The spray, (which was shot backwards before), is now lifted off the surface of the road water and has a ***zero*** velocity ***with respect to the road***. The car body has to go through the spray generated by its own wheels, and the following vehicle goes through the preceding vehicle spray. Many vehicles, eg plates 4, 7, 11 and 18 are not designed to go through this 'wet cloud' produced by this process. The design of bonnets completely ignores the spray produced by the preceding vehicle (5.9, 8.6, 10.7.1).

5.3 PROPERTY OF MOTORWAY SPRAY

You will have observed that in high speed motorway driving, the spray:

a) alters direction haphazardly, (ie it is turbulent) because it;
b) has small momentum, (product of mass x velocity), and;
c) is very dirty with little or no water when it hits the windscreen (figure 16).

Hence it is a requirement of the MOT test, for safety reasons, that the windscreen washers work.

5.4 WATER CONTENT OF MOTORWAY SPRAY

The reason why this spray is ***dirty***, is because the road water is broken up into small droplets as the wheel leaves the wet road surface. As the vehicle speed increases, so these droplets must become smaller. A small

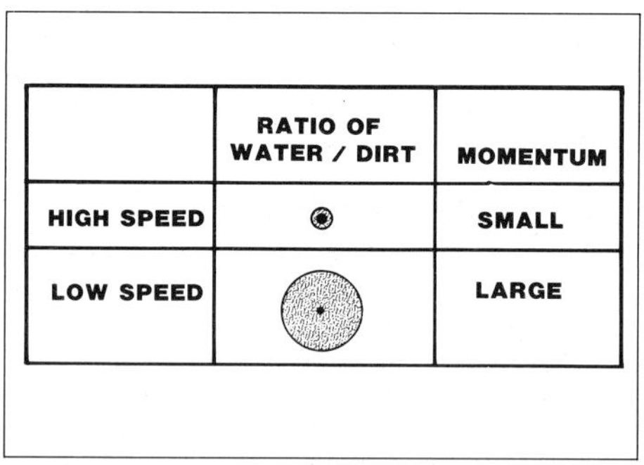

FIGURE 16. A diagrammatic view of the meaning of low momentum particles and the ratio of water/dirt. A water droplet is a small sphere due to the high surface tension of water. This high surface tension makes these small water/tyre dust droplets into highly erosive micro-droplets at high speed which erodes the top paint film(s) down to the base electro-coat on badly designed bonnets (8.6).

water droplet can ***only*** be formed when it has something upon which it can nucleate. An analogy would be the London **'smog'** in the winter before the Clean Air Act became law. Here the water vapour in the cooling air is using the particles of ***sm***oke to form a nucleus for the formation of the f***og***. Because of the high particle content from the smoke, it has become known as ***smog***. In the same way this high speed, dirty road spray, consists of road dirt surrounded by very little water, (figure 16), and so the dirt is not self cleaning or draining and will not wash off the vehicle very easily. The road dirt is a mixture of worn road aggregate, tyre dust etc, and in this form is ***self compacting*** and forms very hard deposits when they dry in places which are out of the direct line of the wheel spray (eg figure 18). At the lower speeds used in the manufacturer's accelerated tests(●), the amount of water which can be associated with each dirt particle increases, and with this added water content, the dirt will not be 'self compacting'. As the wheel rotates slower, the larger water droplets will be thrown off earlier and this gives rise to incorrect designs such as these large holes (plate 6) open to the high speed front wheel spray.

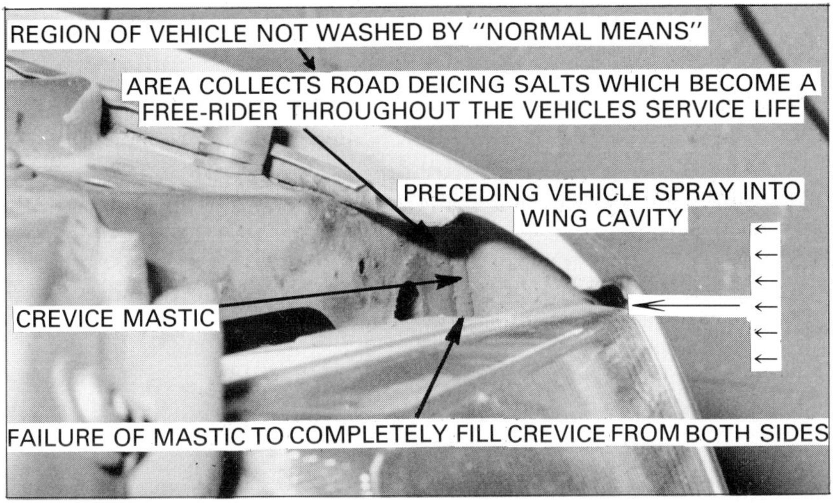

Plate 7. The area shown is the unwashed cavity formed by the headlights and blanking plate which stops mud from forming a mud trap over the rear of the headlights. This cavity now collects the preceding vehicle spray and the road de-icing salts concentrate within the crevices formed by the wing and chassis rail. The photograph also shows the mastic, which seals the wing to the steel blanking plate. Water (carrying the road de-icing salts) is able to collect behind this mastic because it is applied from both sides of the blanking plate thereby forming an 'enclosed crevice'. Mastics also age with time, and this aging can result in the fracturing of the paint film at the metal/paint film/mastic interface. Result: Metal perforation associated with the crevice and mastic. The corroding areas are the oxygen denuded areas. **DIY Rectification** None, but see (10.19). Keep warranty alive and look out for metal perforation along the mastic wing interface.

5.5 DEPOSITION OF NON DRAINING DIRT

The turbulence around the vehicle means that this low momentum spray can be ***deposited in any site*** where it is capable of starting metal 'perforation' corrosion in items which are both subject to the MOT test regulations and come under the 6 year perforation warranty. In some cases the items below may corrode through after the 6 year warranty has expired and therefore early detection, accompanied by where to look and correct interpretation and analysis of the cause of any paint damage is essential.

These are:

- a) the spring hangers, and platforms of suspension units (5.22);
- b) rearward facing panels (5.8) beneath the car to which subframes, brakes pipes* are attached, and
- c) within some sub frame sections eg plate 10, (5.21);
- d) behind both front and rear wheel arch liners figure 14(b);
- e) rear wheel arch returns figure 18, plate 8;
- f) around the petrol tank and associated body work.

* Corrosion of brake pipes: see (8.2).

5.6 NON DRAINING DIRT AROUND WHEEL ARCHES

The problem here is that motorists are not conversant with periodic maintenance and the requirement for washing away the low momentum spray which accumulates over the rear wheel arch lip (figure 18). Furthermore there is some reluctance on the manufacturer's part to give specific winter maintenance instructions, simply because they may give the impression that their product needs such cold and wet 'messy' winter maintenance. Therefore there is no specific mention made to wheel arches within the vehicle owner's handbook. The advice usually given is to wash the car and ensure the drainage holes remain clear. This booklet will come to the opposite conclusion, and suggest that a lot of these drainage holes should be blocked off, or fitted with one way drainage systems. Hosing and steam cleaning behind the wheel arch returns is difficult for two reasons:

1) the steam cleaner spray head cannot get to this area. If it does get in here, it strips away the waxes because it is too near and too powerful;
2) the mud deposited is associated with very little water (figure 16). This has the effect of self compacting the road dirt in this area.

DIY Rectification: Hose away the mud compacts with a fine gentle spray at monthly intervals during winter motoring. Do not use powerful jets to removes this compacted dirt.

Owner rectification of design faults

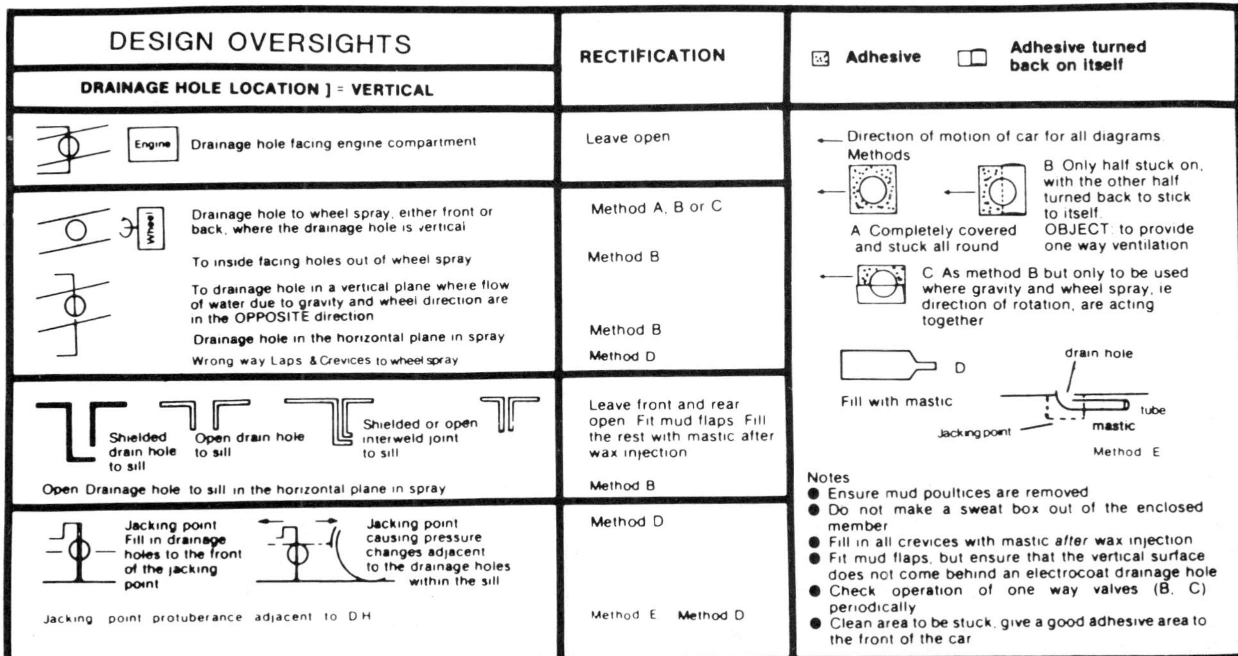

FIGURE 17. Various methods available to the owner for improving durability where there are design over-sights. On no account should the box member/unit be made into a ‘sweat box’ by filling all the process holes. See figure 1 for effect of temperature on corrosion.

Plate 8. The rear wheel arch return shows a large number of wide interweld crevices (between spot welds), which do not contain any waxes. When the vehicle was new these areas contained ***no*** wax, (as shown in plate 9). These bare crevices (of paint, see figure 3, and waxes, see figure 18 and plate 19), are ideal sites for corrosion. The alkali produced by the cathodic reaction of this interweld corrosion is acting as a paint stripper to both the external and ***internal*** paint systems (1.10). Many manufacturers used heat sealing sticks, in order to seal the variable crevice which is formed around this arch. These heat sealing sticks do not allow the waxes to get to the crevice. Hence the provision of the waxes, either by the manufacturer or the after market company, is a waste of time and money if these heat sealing sticks are present. (Heat sealing sticks are present in a large number of makes and models).

5.7 BAD DESIGNS IN THE REAR WHEEL ARCH

Corrosion around both front and rear wheel arches will cause considerable controversy as it is an area where stone chipping could be expected. I shall only discuss the more complicated rear wheel arch structure. The form of this corrosion is shown externally as paint blistering (plate 8) caused by the alkali formed by the cathodic reaction (1.10). The reason for this corrosion could be either stone chipping (very plausible but unlikely), or (and more probable) corrosion induced by the ***perpetual mud pocket*** at this location, but this is not the full story. The 6 year warranty is not a charter for negligence on either party's contractual obligations, and there are problems here which result from the manufacturer's ***'bad design'*** on two accounts. Therefore I would not agree that the owner should bear the whole cost of the repair if this area is not washed regularly. In earlier sections I have stated that a million dollar treatment will not stop the corrosion process if it is subjected to a perpetual mud pocket. The associated crevice corrosion is not fair wear and tear if the following manufacturing design defects are present:

a) the outer panel is often larger than the inner panel, figure 18, plate 8, and therefore I would ask the manufacturer, as a gesture of goodwill, to contribute to the cost of repair (for correct method see (10.19)) on the basis of 50% for the bad design, and owner 50% for lack of maintenance;

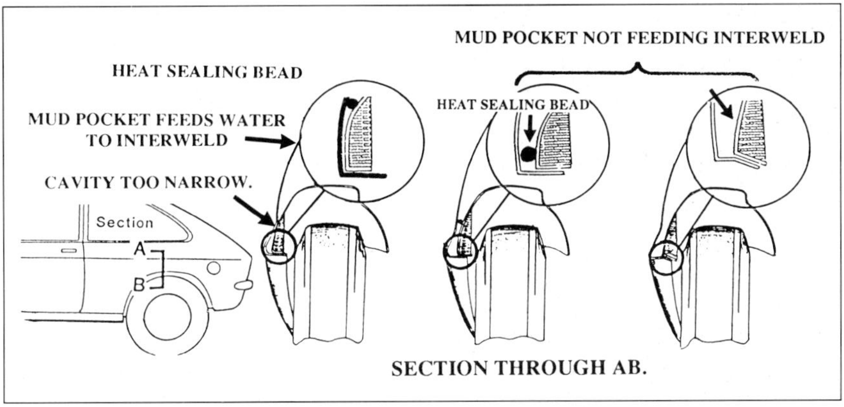

FIGURE 18. The rear wheel arch return is a unique example where no manufacturer has paid attention to design against corrosion. The three sections through the rear wheel arch return **AB** shows: (a) the mud pocket feeding the interweld crevice; (b) a better design but with the heat sealing stick present; (c) the correct design where the road spray does not feed the crevice interweld because the outer panel lip is smaller than the lip on the inner panel. This panel overlap (figure c) should be filled with mastic and the internal panels wax injected. The orientation of the panels will allow easier cleaning. Figure 18 drawn by Maurice Neal and acknowledged with thanks.

b) the above design defect is made worse by the manufacturer because a ***heat sealing stick*** is applied around the rear wheel arch and lies between the two panels (figure 18, plate 9). In this case the waxes sprayed internally to fill the interweld crevice cannot penetrate the crevice due to this heat sealing strip. Corrosion will be initiated in the wax denuded crevices. In this case, age will cause corrosion and the manufacturer's warranty should contribute 85% towards the cost of the repair (method 10.19).

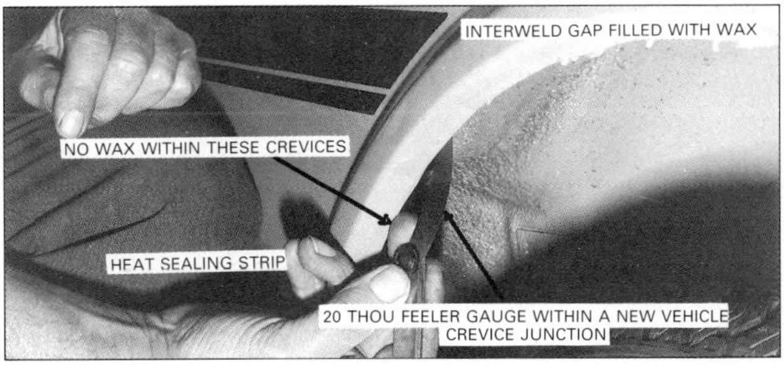

Plate 9. Wax dripping from the crevice in the wheel arch return of a new car indicating a water repellent surface. However further down the crevice, the crevices were devoid of waxes. This could be detected and highlighted for photographic purposes with a feeler gauge. This defect was detected on several new cars. Figure 18 shows the position of the heat sealing strip which keeps the wax out of the crevice.

If the wheel arch area were regularly cleaned, then I would favour the manufacturer paying for the total rectification if either of the above design defects (a) and (b) applied. If you are buying a new or nearly new car, look at the interweld with a mirror as shown in plate 8. On second hand vehicles, the garages will almost always respray this area before putting the vehicle up for resale (10.21).

On new vehicles, the dealers will gladly fill the bare seam with wax from the wheel arch side for you. ***It is easier to do this than to believe that you will make a successful claim later on.***

It is in situations like these that you will find a big difference between the manufacturer's warranty and the after market warranty. The after market warranty will not pay out on this type of crevice corrosion because there is usually an exclusion clause which excludes those areas where the wax could not reach.

The wheel arch returns are not subject to the MOT test regulations (unless corrosion is within 30cm of a spring hanger), even though the arches form an essential safety concept in the unitary body when shunted in the rear, (figure 11d). Wheel arch corrosion, like the front panel to plate 6, will take a lot of haggling because many of the manufacturers will argue that the corrosion is due to stone chipping. I hope I have shown you the truth.

5.8 PRESSURE VARIATIONS AROUND A VEHICLE

If you have followed a heavy goods vehicle on the motorway, you may have noticed the movement of spray in a haphazard and random direction. The spray can alter direction very quickly. This is due to 'pressure eddies' around the vehicle because they are not fully streamlined. Similar pressure eddies exist beneath a car. These eddies take the road dirt and deposit the dirt in out of the way places, eg on the rearward facing panels beneath the car, (just as it is deposited on the glazing of tailgates). These rearward facing members beneath the car are often not coated with either waxes or PVC plastisols, nor are they cleaned at all.

5.9 DENSITY OF TRAFFIC, Overtaking vehicles

During high density traffic, the preceding vehicle causes a turbulent road spray through which other vehicles must pass, (figures 14a and 19). Most vehicle designers omit the influence of other vehicles in their design and prototype testing (8.10), and this leads to the road dirt deposits which now form part of the predictive areas of corrosion, see chapter 10, summary.

This omission by the designers highlights the following problems:

a) incorrectly designed bonnets are eroded by the micro-droplets of

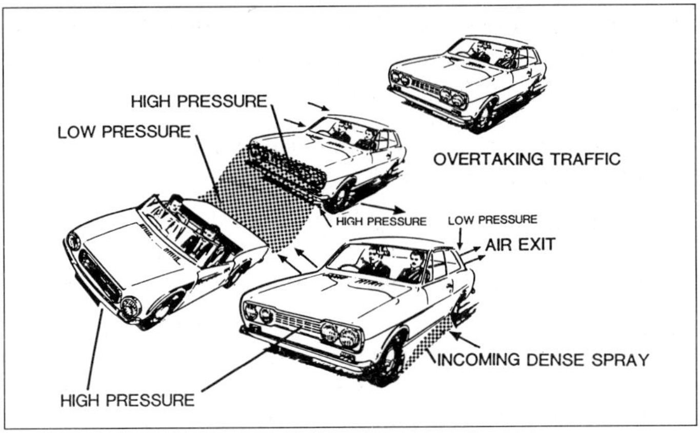

FIGURE 19. A vehicle travelling at speed will produce a high pressure at the front and a low pressure at the rear. An overtaking stream of vehicles causes high pressures on the overtaking side, whilst on the inside there will be a low pressure. The electro-coating conditions in the vehicle body will allow a lot of road spray to enter the drainage holes on the high pressure side. This situation is reversed when the rear of the vehicle comes alongside. The figure also shows that a car in front causes a lot of wheel spray. A following vehicle has to drive through this dense small droplet spray. The motion of the following vehicle results in a high pressure area by virtue of its own speed through the turbulent air. High drainage holes (figure 21C) in the bonnet can often collect the preceding vehicle spray from where it cannot drain. This causes corrosion around the front edge of the bonnet, see also (5.17).

water produced by the preceding vehicle, (figure 19). This erosion is usually down to the base electro-coat, and can usually be seen on the darker colours. It is not due to 'stone chipping', and should be rectified free of charge by the manufacturer within the 18 month* cosmetic warranty period (8.0), (manufacturer's warranty only).

* Some manufacturer's paint warranties are for 12 months, others 3 years (8.6);

b) in the 'box members' of the following vehicles. Overtaking vehicles also cause different forms of pressure eddies, (figure 19).

5.10 TURBULENCE, SIDE IMPACTS

When a vehicle is driven at speed on motorways there is a bow wave of spray coming from the tyre/road interface (figure 14). This is perhaps more noticeable in the slush of winter or if you are a pedestrian. Associated with this bow wave there is considerable turbulence around, and alongside the vehicle. Overtaking vehicles add to this turbulence. This turbulence and the bow wave causes the road dirt to impact the vehicle body at right angles to the vehicle panels. These small particles damage the paint work, and in time they appear as small blisters or rust spots*.

* The 6 year warranty is not a paint warranty (8.6) and does not cover these ***surface nucleated*** items of external paint work damage, see (8.13). It is up to the owner to examine the vehicle thoroughly to see if there is not an alternative reason for the damage, for example the paint damage in plates 1, 6, 8, and 12 were attributed to stone chipping by the warranty agents, (see also plates 15, 13 and 16, figure 25).

5.11 COLLECTION OF WHEEL SPRAY BY THE MANUFACTURER'S PROCESS HOLES

When the vehicle is in motion, say at 60 mph [95km/hr] (88 feet/second), a drainage hole (eg plates 4, 10, figure 20), will collect a large volume of moist air in one hours driving on a wet road. This moist air and dirt is capable of providing water for the corrosion process. Some items are painted separately from the car and added at later stages of the body build (eg wishbones, trailing suspension arms (figure 23), sub frames (plate 10), and other structural parts made from steel sheet). The orientation of these items in their own painting bath is often different to the orientation as to how they are fitted to the vehicle. ***Therefore air admittance and air exit holes (4.2) can often collect the vehicle spray.*** Anybody on the lookout for corrosion (second hand car buyers and MOT inspectors) will be advised to pay particular attention to the process holes (4.2) over the front wheel track. These holes are not for ventilation.

DIY Rectification: Fill the vehicle crevices with waxes or mastic. Fill process holes which face the road wheels; make one way drainage flap valves for larger holes (figure 17).

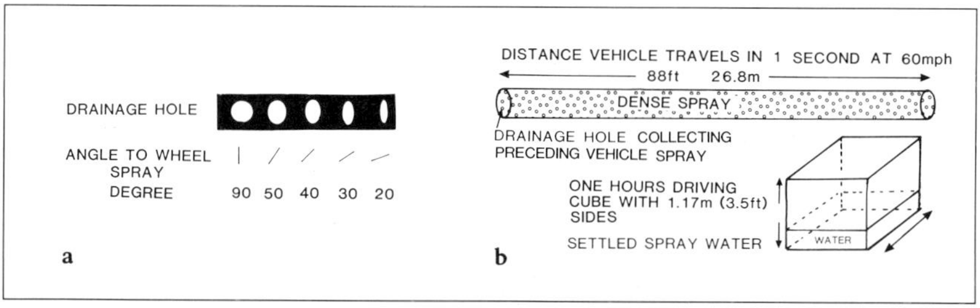

FIGURE 20. (**a**) A manufacturers process hole viewed at right angles to the sheet panel is a circle (plate 4). If this makes an angle to the front wheel spray as shown in plate 4, the apparent area of cross section is reduced as shown here. (**b**) Assume a cross sectional area = $1cm^2$. The cross sectional area presented to the wheel is reduced by ten fold, but, over an hours driving a considerable amount of spray is collected. At a steady speed of 60 mph, 88 ft/second, 26.8 m/second, the drainage hole will collect 1.61 cubic metres of wheel spray, which is equivalent to a cube with 1.17m (3.5 feet) sides. Hence the total amount of "settled" water can be some hundred fold greater than the rainfall. This is because the rain has been recycled by the vehicle wheels.

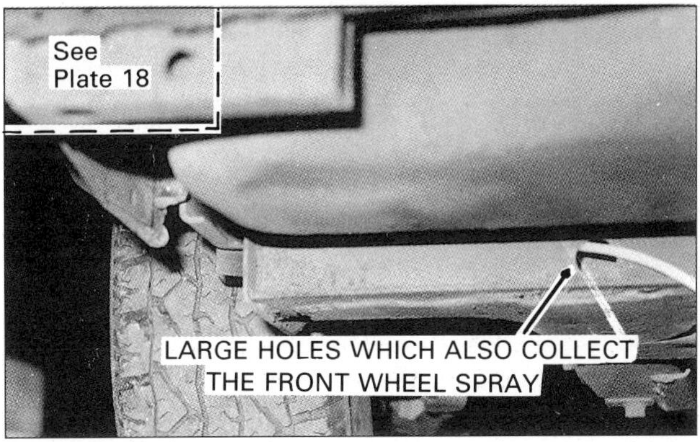

Plate 10. This plate shows the large hole which collects the front wheel spray by virtue of the forward motion of the vehicle. This collected spray resides as a mud poultice at the bottom of the sub frame member where it will nucleate corrosion with the possibility of rust penetration in localised areas. This sub frame member also contains a wrong way overlap (figure 22), which will also collect the vehicle's wheel spray. In a large number of vehicles, the jacking point jettisons water into the sill area through unfilled crevices or drainage holes (see for example plates 17 and 18). **DIY Rectification:** See 10.7.2.

5.12 PASSENGER COMFORT

Fresh heated air is introduced into the passenger car by a ram effect as a result of the vehicle's forward velocity. The amount of air entering the passenger compartment, can be controlled internally by the occupants. Stale passenger air is continually extracted through vents (figure 14a) located in various parts of the vehicle (rear tailgate, rear body adjacent to the C post or through the doors). The amount of air extracted cannot be controlled at all and is dependent upon the vehicle speed. Hence if the admittance of fresh air is reduced at high speeds, the extraction system will still operate as efficiently and lower pressures will be found within the passenger compartment and internal body cavity (box) members. This air continuity path is a requirement of the painting processes as discussed earlier (4.0 to 4.5).

5.13 AIR MOTION IN THE MONOCOQUE BODY

We can therefore ***follow*** the passage of the replacement air (which includes the cloud like road dirt and de-icing salts) as it is ***drawn through*** the sills/rockers (plates 11, and 12), and ***over the rear quarter panels***, (plate 3), depositing the wet road dirt en route. These particles are deposited upon all surfaces, (including vertical surfaces where the ***'adhesive'*** is the dried salt figure 6c), and where they remain for the rest of the vehicle's life. This road dirt cannot drain readily (plate 3) due to this lack of associated water (5.3).

Plate 11. The microclimate of deposited road dirt (adhesively bonded to the vertical surfaces by the crystallised de-icing salts, see figure 6c) can be seen localised around the process (drainage) holes within the outer panel of the sill. This micro-climate becomes a free rider throughout the vehicle's life absorbing water vapour on damp days. Proceed to plate 12.

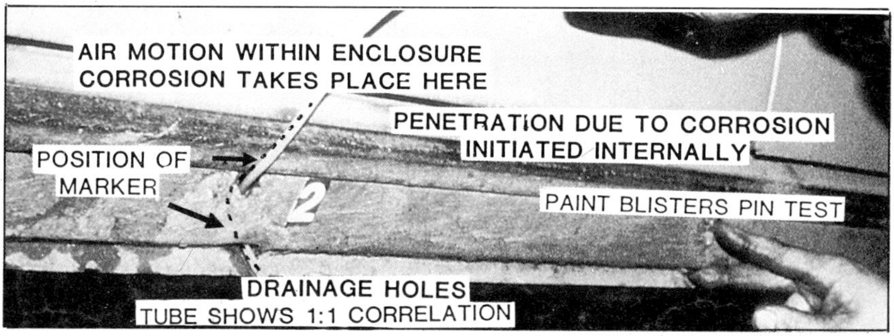

Plate 12. These paint 'blisters' on the outer surface of the sill have been punctured and enlarged with a screwdriver (figure 25) and a plastic "marker" inserted into the rust hole and associated process drainage hole to show the 1:1 relationship between the drainage hole and rust penetration resulting from internally initiated corrosion. **DIY Rectification:** See 10.7.3, figure 25.

5.14 DENSEST SPRAY

The air drawn through the doors and sills by this mechanism, originates from the regions of densest spray as the process holes to the doors and sills are directly over the front wheel track. Therefore, when purchasing a 'second hand' car, pay a great deal of attention to the area around the drainage holes in any panel and its interior whether or not these form part of the stress carrying box members. All panels and bolt on components, (figure 23), are associated with ***process holes,*** air admittance holes (plate 3), electro-coat throw holes (plate 5), and drainage for liquids (plates 1 and 4). (This will be contained in a directive to all MOT examiners to examine these drainage holes in stress bearing 'chassis' members (3.0). Cosmetic panels (3.2) will not be examined (9.1)).

Plate 13. The inside surface of the outer skin of the door panel shows the localised corrosion at the drainage holes, and the corrosion within the hemming fold crevice (figure 3). The suction effect due to the passenger extraction system is exaggerated in those vehicles where the bottom of the doors are in line with the sills or rockers, (figure 21 (b) B). The ***first signs*** of perforation by rust will be at, or around the ***drainage holes*** and about ***10mm (1/2 inch) from the bottom of the outer panel*** (figure 24). These rust through areas will then slowly appear until the whole of the bottom of the door has corroded through. The inside mastic covering the inner hemming fold is to protect the external visual crevice, but it also hides **rust weeps** from the overlap. This mastic will become non adherent by the chemical by-products produced if the design to resist corrosion is wrong, (plate 14, discussed in (5.16)).

5.15 DOORS

The air required to replace the stale passenger air is also drawn up through the drainage holes in the bottom of the doors (plate 13). The associated corrosion is made worse by the design defects discussed below and is not be subject to the MOT test regulations.

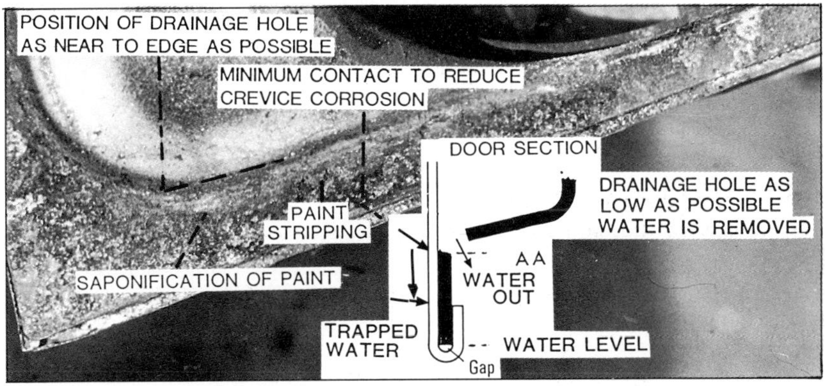

Plate 14. Rain water running down the window glazing can by-pass the rubber at the top of the door (plate 16). This collects in a water trap at the bottom of the door because the drainage hole(s) are too high (figure 21C, plate 15) or not located at the corners (plate 15). This traps water and road de-icing salts in the crevice formed by the clinch fold. On drying these de-icing salts will be concentrated in the clinch fold. The crevice is also the area denuded of paint and oxygen replenishment (figure 3).

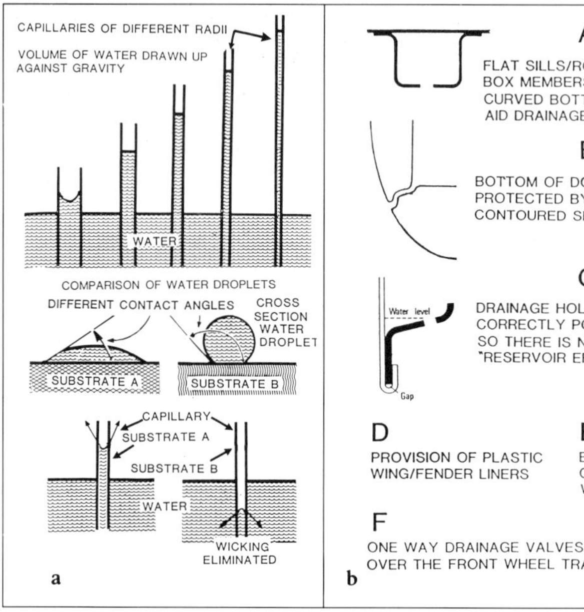

FIGURE 21. (**a**) Capillary wicking is shown diagrammatically and is dealt in more detail in other texts (●). (**b**) summarises and shows the basic 'good design points' which should be incorporated in all vehicle designs. Note; (A) Better drainage. (B) Doors scoop up front wheel spray (plates 4 and 13). (C) The reduction in the amount of metal overlap in (C). (D) Correctly designed wing liners (figure 14). (E and F) Can only be done properly by the manufacturer. The DIY concepts are considered in figure 17. p-35

5.16 BAD DESIGNS OF DOORS

The corrosion associated with doors is accelerated where the doors contain either:

a) a high internal lip which causes water to remain in the internal door crevice for longer periods than is necessary. A cross section of this is shown for clarity, figure 21 b(c);

b) where the drainage holes are not at the extreme corners of the doors, plate 15. Rain water by-passing the glazing rubbers means that the vehicle cannot be garaged on uneven ground without the doors partially filling with water, plate 16;

c) if the metal overlap is too large. This large overlap (figure 21b (C)) creates a capillary wicked reservoir of water for the corrosion process, plates 13, 14 and 15;

d) the bottom of the doors are in line with the rockers (plates 4, 13). The correct design is shown in figure 21(B);

e) no waxes used because one side zinc coated steels are relied upon to give corrosion protection (figure 24);

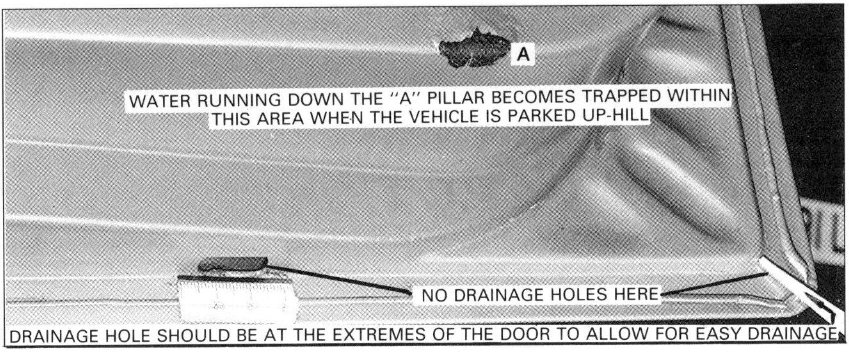

Plate 15. The drainage holes are not at the corners of the doors, as they should be for 'good designs'. If this vehicle is parked up-hill, the rear of this door will fill with water and cause damage within the door. Internal corrosion of the door panel is expected within the six year warranty period. The corrosion shown at point A, results from a design defect on the 'A' pillar, which allows water to become trapped behind the door rubbers when the vehicle is parked up-hill. This should be repaired under the 6 year manufacturer's warranty free of charge. (You will be told that the defect arises from the seat belt getting caught in the door. You will not find any other marks on the 'B' pillar to substantiate this counter allegation).

f) incorrect door glazing rubbers. As a good example to photograph refer to plate 16. There are several incorrect designs of glazing rubbers on the market and a detailed examination of the rubbers with the windows down, will show that many rubbers channel the rain water into the door enclosure;

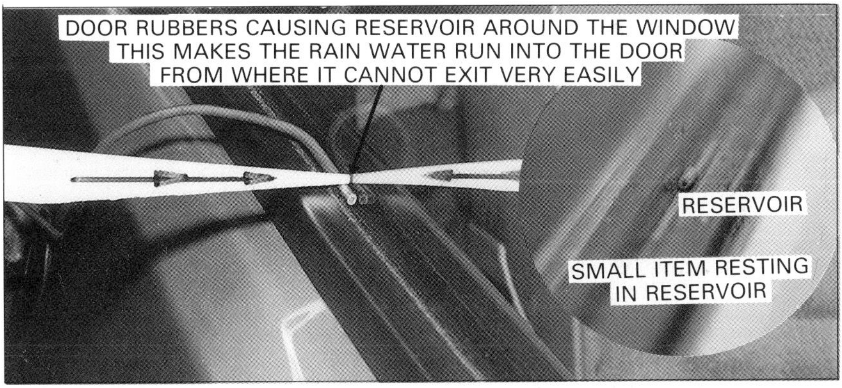

Plate 16. Shows the reservoir produced by the 'non water shedding door rubbers'. To highlight this area for photography, a small piece of wire is resting in the 'reservoir' and against the glass. Rain water is therefore forced to run down into the doors where the drainage holes are not ideally situated. Other incorrect designs of glazing rubbers on other vehicle manufacturer's products do not photograph well, because the detail is small.

g) door handles channel rain water into the door enclosure when parked on uneven slopes.

DIY Rectification There is no rectification possible on the photographed glazing rubbers which are common to several manufacturers. Vehicles with these defects would not be an ideal choice for a second hand buyer without a garage. Some glazing rubbers require a mastic end stop to the rubber glazing strip to stop water running into the door. If you had to buy such a vehicle, it would be advisable to remove the internal door trim* to inspect the internal crevice.

* Usually a ten minute job with a screwdriver to pop out the fixings. Some vehicles do require a special circlip tool to remove the window winder arm to allow the internal trim to be removed.

5.17 CARS GARAGED OUT OVERNIGHT ON SLOPES

The bad design of the doors on so many vehicles means that you should be warned about leaving the vehicle out every night. The rain water which falls onto the roof and bonnet has to be drained away. For good design against corrosion, these drainage channels should not contain any crevices. Lack of paint and waxes within the crevice, (and holidays* in the mastic covering crevices), means that crevice corrosion may take place accompanied by the stripping of the surrounding paint film.

DIY Rectification: Use appropriate method to keep water out. (*A holiday within the mastic will allow water to be capillary wicked throughout the gutter system. This is particularly important on the 'A' pillar due to the high pressures produced here as a result of the vehicle's forward motion).

Those 'A' pillars with "streamlined" (cosmetic trim) covers may trap small leaves and dirt adjacent to the 'A' pillar causing corrosion behind the enclosure in an area which is out of sight. This should be covered by the warranty as it is bad design.

DIY Rectification: Pass small plastic feeler gauge from the bottom to the top to remove all trapped material.

Streamlining by producing gutterless 'A' pillars can allow water to become trapped behind the door rubbers as shown in plate 15.

DIY Rectification: Tape PVC electrical insulation tape adjacent to door rubber, or dry area after rain.

These points can be further supplemented by considering the accumulation of leaves and other debris which can enter the heater grill and become lodged behind the plastic wing liners where they block up any drainage provision.

As an exercise, follow the rain water running off the roof, windscreen and bonnet for different parking slopes for your present vehicle. Consider the vehicle then subjected to a 60 mph wind from the front. Could any of this water run into the following areas:

a) windscreen glazing rubbers; where does the windscreen wiper deposit the road de-icing salts in the winter?;
b) spot welded metal overlaps to the front of the vehicle;
c) bonnet stiffeners and crevices through the heater air intake;
d) into crevices formed by the 'A' pillar and the engine/passenger bulkhead;
e) feed the wing/chassis rail through wrongly sized washers and bolts;
f) into the longitudinal chassis members at the various intersections of the panels;
g) petrol flaps and their (crevice) recesses and then subsequently drain away to finish up within the rear wheel arch;
h) box and chassis members. The forward motion of the vehicle can force capillary held water into these members;
i) can the tailgate be opened on a transverse slope without water draining into the glazing rubbers;
j) enter the passenger extraction vents. One manufacturer allowed rain and melting snow to enter the rear wheel arch area via this mechanism;
k) reside around the front or rear light clusters and can therefore make contact with either a horizontal surface or feed a metal overlap;
l) if the panels contain holes for additional equipment, eg wiper motors etc, can they form water reservoirs thus keeping the crevice perpetually wet (1.10).

DIY Rectification: Use appropriate method to keep water out. Coat behind glazing rubbers, light clusters etc with solvent based waxes.

You will quickly see that the door glazing rubbers is not the only problem.

5.18 EXTERNAL MEMBERS – PROTUBERANCES

Any protruding member, (figure 15. plates 17 and 18), at right angles to the direction of motion will generate ***high pressures in this locality*** thus deflecting the vehicle's front (or rear) wheel spray, (and/or the preceding vehicle spray) into the box members, (plate 17). Hence jacking points, slung beneath the sills, and adjacent to the process drainage holes will be susceptible to corrosion and will warrant close inspection. The jacking point, (shown in plate 18), effectively compresses the front wheel spray producing a high pressure by virtue of the vehicle's own motion. This jettisons the road spray into the sill member through "holidays" within the PVC plastisol sealer. This will lead in time to internally nucleated

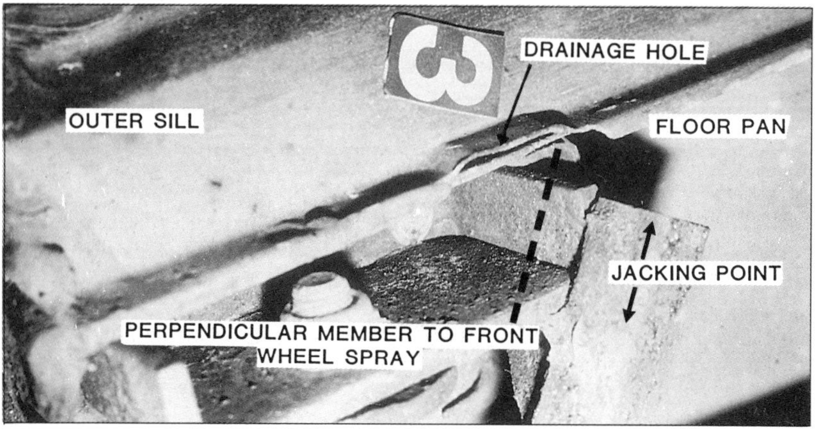

Plate 17. This plate shows the jacking points adjacent to a drainage hole. The jacking point will deflect the (front) wheel spray into the box members. The same design faults occur with other protuberances: mudflaps; cosmetic guards which cover outward flared wheel arches; doors not aligned with the 'C' post; internal wheel arch stone guards. **DIY Rectification:** See 10.14.

corrosion. This high pressure generated by the jacking point will reinforce the normal suction of the spray through the sill/rocker box members by virtue of the passenger air extraction system (5.12).

Other items to look out for are:

a) plastic rear wheel arch liners projecting beneath the sill or rocker members;

b) interweld seams, or "ventilated" (8.10) box sections giving a perpendicular member to the front wheel spray, (eg plate 18).

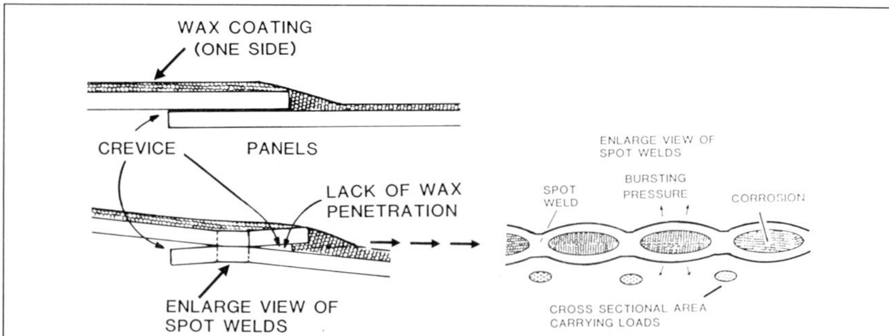

FIGURE 22. Thick viscous waxes (or PVC plastisols) will bridge the crevice of a metal interweld or overlap. If the other side is uncoated, water and salts may get into the crevice and thereby concentrate the de-icing salts within the crevice. The fact that one side is coated and bridged, means that the crevice remains wetter for longer as drying is restricted. For interwelds, a highly penetrating oil or wax must be used (6.3-6.4).

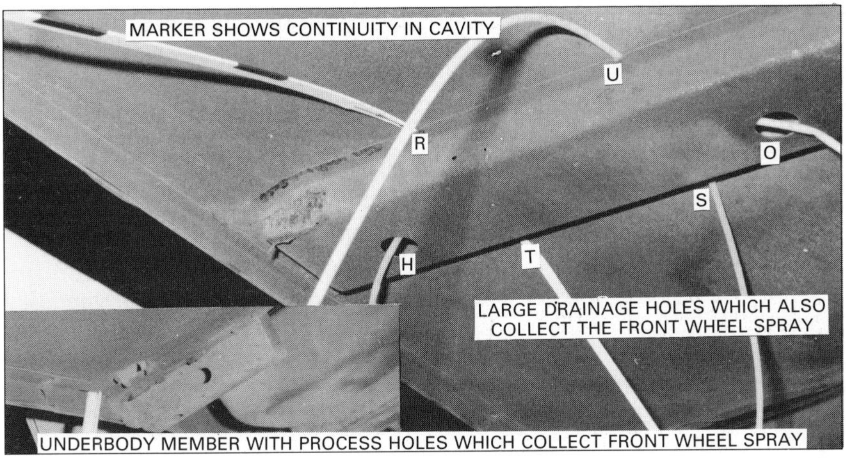

Plate 18. View towards off side rear wheel. This plate shows the underbody member which collects the vehicle's front wheel spray through the electro-coat air admittance holes 'R' and 'U' (4.2(b)). Larger holes 'H' and 'O' (4.5) will also collect the front wheel spray, and the preceding vehicle's spray due to the low pressure developed, by virtue of the holes highlighted as 'T' and 'S' which are to the rear of the box member (4.2(b)). Why the two sets of air admittance holes 'R' and 'U', 'T' and 'S'?. If the car enters the electro-coat bath front first (plate 2), the holes 'T' and 'S' will act as the air exit holes for the paint entering the member. The car will come out of the bath front first, and the holes 'R' and 'U' will act as air entry holes so that the paint could run out evenly. Note the angle of entry in plate 2. **DIY Rectification** of this bad design detail can easily be undertaken by filling in holes R and U, and provide a one way flap valve (figure 17) to holes H and O. There are better ways for the manufacturer to rectify this defect. As to why the manufacturer does not rectify them, see (8.10).

The inset photograph shows the interweld gap adjacent to the jacking point. The front wheel spray will enter this gap and cause internally initiated corrosion. As the outer panel is outward flared, the first signs of rust perforation will be dismissed as due to stone chipping. **DIY Rectification**: Fill with mastic or thick waxes, keep warranty alive, and test all paint blemishes as shown in figure 25.

c) any protuberance or item which is housed underneath the vehicle body (eg petrol tank), or box members which constrict the air motion behind the front (or rear) wheels is capable of forcing the road spray into process holes.

5.19 WRONG WAY OVERLAPS

Wrong way overlaps (figure 22), will:

a) collect the preceding vehicle spray, or act as a water reservoir;
b) increase in pressure at the leading edge of the overlap which will

c) force the road water into these overlap crevices. Any road de-icing salts will concentrate within them when the area dries;

d) give rise to erosion of the paint film at the leading edge due to the abrasive effect of the low momentum spray, figure 16.

DIY Rectification: See (8.6, and 10.19), claim upon the manufacturer's warranty. If rejected, enter a small claims action against the retailer, see (10.22).

5.20 FORCE TRANSMITTING MEMBERS

Wishbones, trailing suspension arms, sub frames, and other structural parts (plate 18) are made from steel sheet to give a 'partially open box' or cylinder, (5.11, 5.14). A pressure difference will develop between the open ends, because the ends are exposed to different air speeds, (figure 23). One open end, (fixed to the car body) is partially protected by the front air dam, whilst the other open end, (fixed to the axle) is in a faster

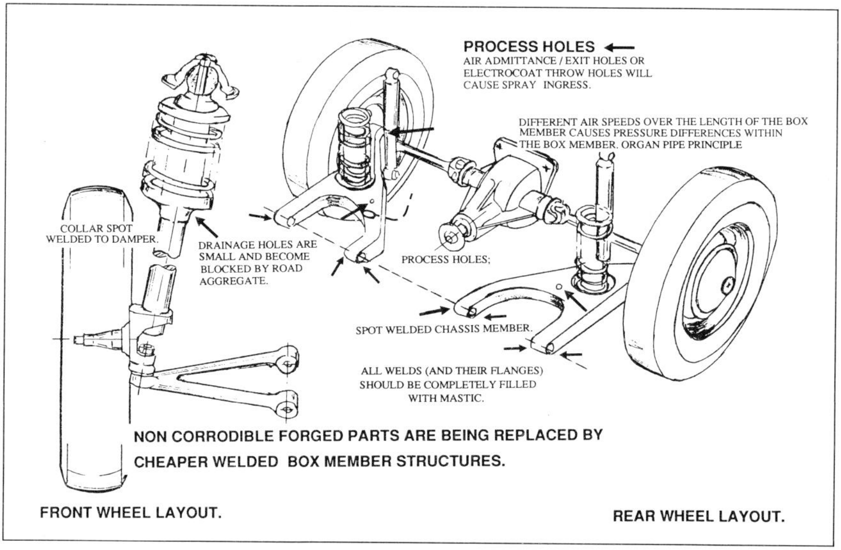

FIGURE 23. Suspension components are made from box membered structures because they are lighter and cheaper than forged parts. They are often painted in a separate paint bath by either dip primers or by electro-coating. Their orientation on the vehicle is often different to the orientation in the paint bath. Air admittance and exit holes (4.2) still have to be provided and these can collect the wheel spray. **DIY Rectification**: These holes are indicated and should be filled with mastic, (figure 17), because they will be adjacent to a perpendicular member, (eg their fixing point). Road stones (chippings) are capable of blocking drainage holes in the suspension collars to act rather like a dam. Figure origination and acknowledgement: Automobile Association, How the car works, pages 145 and 147. Figure 23 redrawn by Maurice Neal and acknowledged with thanks.

moving air steam. The resulting pressure difference will allow the open end to collect the wheel spray of the preceding vehicle and deposit it within the member.

5.21 TRAILING ARMS

Trailing arm suspension units (figure 23) do have ***process holes*** which collect (figure 20) the front wheel spray and/or the preceding vehicle spray. These holes, often 1.5″ in diameter, should be protected by plastic grommets (figure 17) or taped over.

5.22 THE COLLARS ON MCPHERSON STRUTS

The interweld of the lower collar to the strut containing the hydraulic damper is an important part of the vehicle's load bearing structure. This area is subject to the non draining road spray and the drainage hole in the lower collar often becomes blocked due to loose road stones.

DIY Rectification These drainage holes should be checked periodically after the summer "tar and stone" resurface treatments. Corrosion within this lower collar suspension unit will lead to the collapse of the suspension unit. Due to the thickness of the collar, it is often difficult to detect a problem within the interweld, but a blocked drainage hole should give an early warning to the likely problem.

5.23 SUMMARY

See 10.4 onwards.

CHAPTER 6

WAX INJECTION

6.0 THE BEST PROTECTION IS *GOOD DESIGN*

The protection of any steel article is best achieved by good designs to resist corrosion. Wax injection came about because these principles in design were not being achieved.

6.1 PROTECTION OF 'CREVICES' and INTERWELDS

The interwelds can be protected by several methods. Historically there has been three methods:

1) ventilation of the enclosed member;
2) wax injection (plate 19). The capillary creep of the waxes will be discussed in the next section;
3) precoated steels. These are discussed in Chapter 7.0.

All three techniques form the basis for the manufacturers 6 year corrosion warranty. Up to now, I have shown that ventilation does not work, in fact, it makes the task of protection even harder.

6.2 WAX INJECTION

The after market rustproofing companies started the wax injection programme of cavity members (see 8.0). This technique has been imitated

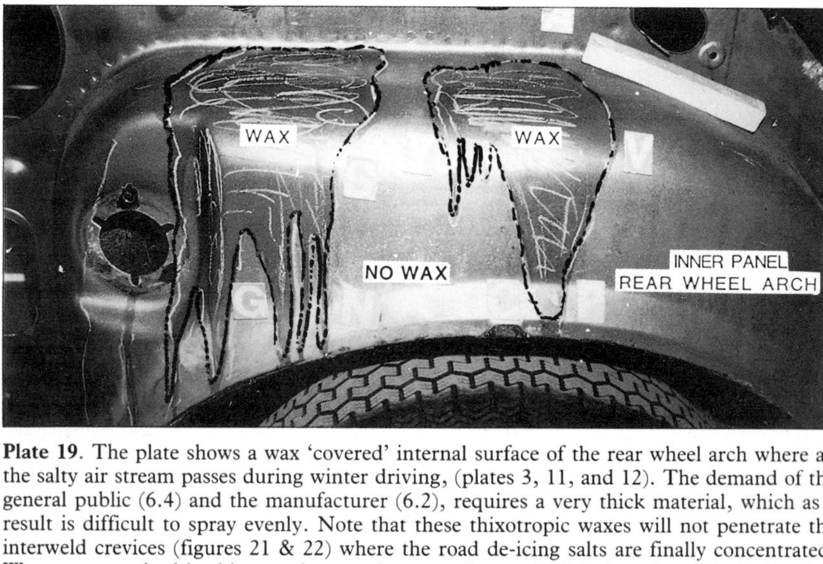

Plate 19. The plate shows a wax 'covered' internal surface of the rear wheel arch where all the salty air stream passes during winter driving, (plates 3, 11, and 12). The demand of the general public (6.4) and the manufacturer (6.2), requires a very thick material, which as a result is difficult to spray evenly. Note that these thixotropic waxes will not penetrate the interweld crevices (figures 21 & 22) where the road de-icing salts are finally concentrated. Waxes are required in this area due to other contributory bad designs (see 1.5, 5.17(g)(j)).

by the motor manufacturers in order to gain ***added*** protection on ***selected*** corrosion prone internal box sections. The waxes are sprayed into the cavities to add further protection to the thin electrocoat within both internal box members and their crevices, (figures 2, 3 and 11, (4.4 to 4.8)). The thickness of wax sprayed on is 25 to 100 microns. The uniform distribution of waxes is often not achieved satisfactorily as shown in plate 19 as a direct result of other requirements made of the waxes used. These other requirements have nothing to do with their corrosion protection properties. The most important requirement made by the manufacturer is that the waxes must not drip. This arises due to the wax injection taking place before the vehicle has been to the trim shop, and therefore if any dripping did take place, it would make the factory floor and working conditions unpleasant. Hence the waxes are engineered so that they 'freeze' in the absence of high shearing forces. This makes the waxes ***difficult to spray*** (plate 19) and they do not penetrate crevices very well (figure 22).

6.3 PROPERTIES OF WAXES REQUIRED FOR CORROSION PROTECTION

The penetration of the waxes into the narrow (capillary) interwelds should be the first requirement for corrosion protection from the wax injection system. This arises because:

a) the spot welded crevice (interweld) is an area which is naturally denuded of any oxygen replenishment (figures 2 and 3) and is therefore ***more prone*** to corrosion (1.6). This corrosion is accelerated because these areas become ***acidic*** in nature (1.10) and suffer preferential attack;

b) the interweld is denuded of electro-paint (figure 3), (4.8). If the interweld gap is made smaller, there is less paint deposited and the more efficient is the corrosion reaction, and the more acidic is the local micro-climate;

c) the road de-icing salts concentrate in the liquid which remains when the winter slush dries by evaporation. The last area to dry will be the interwelds (and mud pockets). Hence the road de-icing salts will concentrate within the interwelds (and mud pockets).

6.4 CAPILLARY PENETRATION OF WAXES

The waxes work because they hate water and water hates the oily surface. Therefore water 'balls up' on the wax surface (figure 21a). A crevice with a waxy surface will exclude the road de-icing solution from the interweld even though a capillary crevice is still present. This also means that the road de-icing salts will not be concentrated within the interweld capillaries. The ***thickness*** of the wax coating does ***not*** play a part in this protection mechanism. The most important point to appreciate in this discussion is the fluids must be ***thin enough*** in order to ***penetrate*** the narrow capillaries to ***exclude*** water. Water has the best penetrating properties known to man, and so will gain access by wicking into any capillary not coated by the waxes, and produce crevice corrosion with the resultant high acidity (1.10). However, thin liquids will easily drain off a vertical surface. Where the wax is giving an extra thickness layer to the electro-deposited paint to give added corrosion resistance to the deposited salt particles, thickness and coherence will be important. To produce a ***thick*** film from a liquid on a vertical surface requires a liquid with a ***high viscosity***. Thus there are two opposing requirements as regards the viscosity of the waxes used. (See also exclusion clauses 8.3).

The AA (Automobile Association) ran an ***"AA Approved"*** system on behalf of its members to approve the after market wax injection companies. This 'approval' centred on giving approval to the waxes used by the rustproofers, rather than the technical aspects of application, (plate 19). The AA's tests used salt sprays. The performance of all protective coatings, like waxes, increases with film thickness (2.5). To obtain ***thick*** films on ***vertical*** surfaces necessitated a ***high viscosity*** wax. (The customer also wanted to see a thick wax film for his money). As the application of the wax was usually when the vehicle had been steam cleaned, the addition of water displacement additives (●) was also considered a necessity. This combination meant that the steam cleaning water which by-passed the door glazing rubbers (plate 16) became trapped in the lower door clinch fold (plate 15). The trapped water, (from the steam cleaning process), excluded the waxes from the very crevices which were requiring protection. Furthermore, in order to measure capillary penetration with thick waxes, the AA tests were gravity aided and the waxes were placed on top of the crevice. Unfortunately the AA Engineers did not lay sufficient stress on capillary ingress ***against*** the forces of gravity (figure 21a). This gave rise to the after market wax injection companies going for greater sprayed thickness to achieve AA approval for longer and longer times in the salt spray tests. The result was that these thick waxes bridged the crevices (figure 22) and early failure resulted. These approval tests should have made a correlation between the time in the salt spray with the ***coating weight per unit area*** together with the ability of capillary wicking against gravity. Penetration into the capillaries should be of paramount importance because crevices form ideal corrosion batteries when wet.

6.5 WAXES AND THEIR SOLVENTS

Other parameters to be considered are:

1) the waxes are 'suspended' or 'dissolved' in the solvents. These 'solids' may dam up the inlet to the capillary and block the capillary interweld;

2) the solvents may rise up the capillary and it is quite possible for that solvent to contain no wax. Hence the ingress of the solvents into the capillary will not take any wax with them, so that when the solvents evaporate, there would be no protection to the 'unpainted' interwelds (figure 3).

6.6 WAXES USED

Most waxes used in the present series of wax injection programmes undergo a transformation upon heating and drying which manifests itself as a ***'mud crazing'*** effect, where cracks appear within the wax films (especially where pooling or the thickness is greater than normal). These cracks often appear where the wax has been allowed to pool at the bottom of doors and within the sills. These cracks effectively form another crevice within the interweld and allow the road salts to concentrate within them. Where retreatment is specified e.g. as accident repair procedures, or as a condition of keeping the warranty alive (8.12 [1] [2]), there is a danger of trapping road de-icing salts within these fractured areas.

6.7 SUMMARY

Waxes work because water hates a waxy surface. However to work properly the waxes must be thin enough to enter the crevice to exclude the water which will give rise to corrosion. You cannot wax inject a wet car because the surface tension forces of water are much higher than those of the solvents in which the waxes are carried. A water displacing agent (•) added to the wax will only displace the water to the very crevices which you are wishing to protect. Wax injecting internal panels can trap de-icing salts in crevices formed by defects in the old wax, (see 8.12, conditions [1] to [6]).

CHAPTER 7

PRECOATED STEELS

7.0 GALVANISATION

Galvanising describes steel which is coated by immersion in ***molten*** zinc to form the coating. The steels used in the automotive industry are electro-galvanised, where the zinc on these steels is electro-deposited, and called electro-galvanised (EGL) steel. Electro-galvanisation is often shortened in the motoring literature to ***galvanisation.*** The steel plate can be coated on one side or both sides depending upon the configuration of the electro-deposition. This process is expensive and only ***very thin*** coatings of zinc are deposited. The steel is given a numerical code eg. **CR25** where CR = **c**old **r**olled, and the number (**25**) represents the thickness of zinc deposited in microns. [1 micron is 10^{-6} metre would be 0.04 thou]. The coating on galvanised steel is fairly thick. The term 'galvanised' steel has been borrowed to ***market the concept of protection***, because galvanisation has a proven product reliability.

7.1 WHY ZINC?. CORROSION PROPERTIES OF ZINC

In general zinc is a very good protection system for steel, because it not only forms an adherent protective film on top of the steel, but it corrodes sacrificially (ie preferentially) to the steel. The corrosion of zinc is acceptable to society because the corrosion product is not coloured like that of steel and so one cannot 'tell' that the zinc is corroding. This acceptance of corrosion, provided ***colourless by-products*** are produced, means that zinc can be sacrificed to protect steel. This colourless corrosion product also forms a coherent protective layer of a sparingly soluble salt (the corrosion product) on the zinc and so protects the zinc from further rapid attack. This form of protection is used, for example in: old domestic galvanised window frames, motorway crash barriers, galvanised pedestrian railings, and ships at sea, plate 20.

7.2 PHYSICAL PROPERTIES OF ZINC

Zinc has a low melting (419.5C) and low boiling point (907C). Zinc is therefore capable of being volatilised from the spot welded area, leaving a small area within the interweld with little or no zinc coating. The comparable melting and boiling points for iron (1535C, 2800C) are given for comparison. Thus the loss of zinc during welding, from oxygen depleted crevices (figure 3), is a strong argument for wax injecting the zinc coated steels.

The benefits of zinc can best be described by looking at how the metal is used in the marine industry.

Plate 20. A ship in dry dock. This shows the sacrificial anodes on both rudder and hull, which are replaced at regular intervals. These zinc anodes prolong the life of the steel hull in sea water. The zinc is used to protect the ship's hull when the paint film is damaged by floating debris. In this case the zinc is a reinforcement protection system for holidays in the paint system. I have chosen this plate because it shows six important points; (**1**) a (corrosion) current must flow in the ***external liquid*** environment between the anodic (dissolving zinc) and cathodic sites (holidays in the paint); (**2**) the ***whole mass of zinc*** is available for sacrificial use in protecting the steel hull; (**3**) the zinc does not have to be a ***continuous pore free coating*** to achieve protection of the steel; (**4**) the zinc will not protect the superstructure of the boat (unless it sinks); (**5**) the zinc is not painted, see also (7.9), and (**6**) the ship's hull has to be immersed in a good electrolyte.

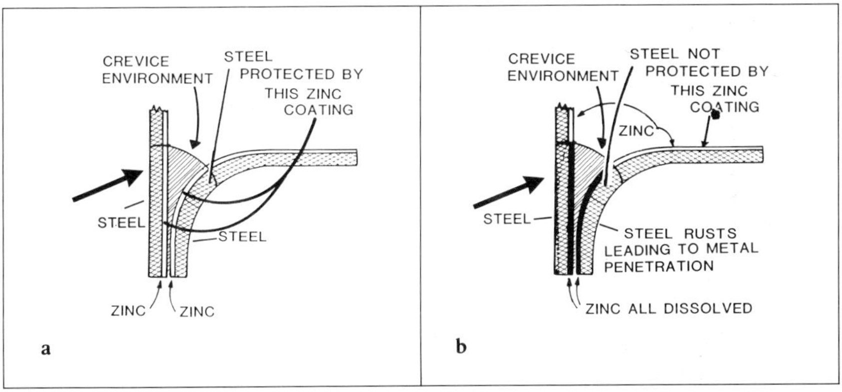

FIGURE 24. The meniscus formed along the metal/air/water interface will form an ideal corrosion cell, (figure 3). (**a**) One side zinc coated steels face the internal panels of vehicle bodies and the zinc protects the adjacent metal if there is good aqueous contact. (**b**) Once the zinc has dissolved*, then the zinc which is not in the aqueous environment cannot offer (sacrificial) protection. The arrows show the correct orientation of the jeweller's screwdriver (figure 25) where double skins, clinch folds, interwelds etc are suspected.

* Similarly with galvanised window frames when the zinc is used up in the glass/zinc interface, the steel will corrode. The external zinc, which is out of the liquid environment will not act sacrificially.

7.3 CATHODIC PROTECTION. SACRIFICIAL METALS

In the ship's case, zinc or magnesium slugs are screwed to the hull of the ship below the water-line and these dissolve (plate 20). In dissolving they protect the hull of the ship from corrosion. These slugs are called sacrificial anodes and the form of protection is called 'cathodic protection'. The success of this form of protection is the reason why the motor industry has gone over to galvanisation using precoated steels.

7.4 ZINC AND THE MOTOR CAR

Zinc does not work in the same way on the car as it does on the ship's hull. This is because to stop motor vehicle corrosion, the vehicle must be kept dry. Hence the zinc ***cannot*** protect the steel if it is: **(1)** ***out*** of the liquid environment, (figure 24) or, **(2)** where the same liquid environment does not make contact with ***both*** the zinc and steel, (discussed in the next section).

7.5 ZINC IN WINDSCREEN WASHER BOTTLES

I am often asked the question:

"Why can't you protect the car by putting a lump of zinc into the windscreen water washer bottle or under the rear wheel arch?"

The secret to understanding this is in the concept of the total circular corrosion current shown in figure 4. The metal dissolving produces a negative particle, which we call the electron, together with a positive ion (figure 4) pertaining to the corroding metal in question. The corrosion current must go round a complete loop. Figures 4 and 6(d) shows the electron current within the metal. At the cathodic site, the negative charge of the electron is changed into the negative ion. This negative ion then goes off in the liquid environment to react with the positive ion produced at the anodic site and, if the metal is steel, rust is formed. The ***rust*** formed, is a ***new compound*** resulting from the ***two ions*** (that produced at the anode, and that produced at the cathodic site). If this combination of the two ions did ***not*** take place there would be ***no*** corrosion (at the anodic site). The paint film works by stopping this current, (figure 8). If the zinc were in the windscreen washer bottle, it could only corrode with the zinc being host to both the anodic and cathodic reactions.

If zinc is going to protect steel, we must make it very difficult for the cathodic reaction to take place on the zinc but very easy for the cathodic reaction to take place on the steel. The zinc must make metallic contact with the steel, (which is very easily achieved), and both zinc and steel must be connected through the ***same*** liquid environment, (which is less easy to achieve). When the zinc corrodes, the electrons produced by the dissolving zinc travel through the metallic contact and are changed into negative ions at the steel surface. These negative ions, produced at the steel surface, must now go back through the liquid environment to form a new zinc compound with the dissolved zinc ions. As these zinc ions are in the washer bottle, we would have to provide a wet path from washer bottle to car body. As this cannot happen ***easily***, there will be no sacrificial protection. This series of events does take place for a ship at sea (plate 20) but not for the structure of the ship which is out of the sea.

7.6 CATHODIC PROTECTION BY IMPRESSED CURRENT

The sacrificial metals used in the shipping industry are ***not*** coatings (plate 20). However they do protect the whole structure below the water-line because the ship is totally surrounded by a lot of water. A similar flow of current which is obtained from the zinc corroding, can be obtained using impressed current techniques (•) (ie from an externally applied d.c. source (1.5)).

This is an alternative to the metal slugs illustrated in plate 20. It saves the expenditure on dry dock fees as the anodes do not require replacing. Here again the impressed cathodic current does not protect the superstructure. The current within the electrolyte, (the sea for ships), can spread out from the inert impressed current anode and protect the whole of the ship, (which is immersed in the sea). Certain devices using these impressed current principles running off the 12V car battery, are marketed world wide for use on the motor car but they do not work to provide protection on the motor car because there ***is no*** current flowing in an ***external liquid*** environment. The physical requirements to stop motor vehicle corrosion is to ***get rid*** of the water environment, but this external liquid environment of water is a ***requirement*** for cathodic protection ***to work***. Thus the two mechanisms are incompatible.

7.7 ABSORBED WATER ONTO PAINT FILMS

Neither process, cathodic protection by impressed currents, or sacrificial zinc will protect the whole car like it does with the ship because of the high resistance of the thin film of water absorbed onto the vehicle's paint film. As a result of this high resistance, the flow of the ionic current (figure 4) will be very small, and so there will be ***no*** impressed current. This high resistance was the reason why a paint film stopped corrosion (figure 8).

If this argument does not convince you, it may help to consider where these anodes are placed on the car: (1) inside the boot and (2) within the engine compartment. In both locations there is very little moisture present. If they were put beneath the wheel arches, the alkali produced (1.10) would strip off the vehicle's paint system.

Similarly zinc coated steels will only protect a small area of the interweld from the aggressive micro-climate and not the whole structure or panel which may be zinc coated (Figure 24). Zinc coated steels only protect the steel at the crevice, ***and only*** if the zinc makes contact with the same environment which is in contact with the steel.

7.8 ZINC AND OUR SURROUNDINGS

Motorway crash barriers show marked localised corrosion due to 'salt deposits' where the rain has not washed the road de-icing spray off the barriers. There is plenty of zinc still available on the barrier, but it cannot be used sacrificially because it does not form a liquid contact with the corroding steel. In the automobile accelerated salt spray test (•), this zinc would be capable of being used to protect the barrier because the test samples are very wet and covered with a semi-continuous film of water. This is another example where the industries accelerated tests do not simulate the correct environment when the vehicle is out on the roads, (8.10). The automotive impressed current cathodic protection system was tested using the salt spray test and it showed promise for similar reasons.

7.9 LACK OF PAINT ADHESION TO ZINC PRODUCTS

The disadvantage with using zinc coated steels and then painting the zinc is shown by the following galvanised structures: pedestrian railings; domestic galvanised window frames. In these examples it is very difficult to achieve ***paint adherence*** to the zinc. A dry paint brush will "tickle off" the paint film from a galvanised window frame.

7.10 METALS IN DIFFERENT ENVIRONMENTS

Different metals behave differently in the same environment, for example; steel only dissolves in neutral and acidic environments, whilst zinc dissolves throughout all pH (•) ranges. Thus zinc corrodes in both the acidic anodic sites and at the alkaline cathodic sites and this is why paints often lack adherence to zinc products. Hence in moist environments, (eg interweld and flange crevices), the metal zinc effectively dissolves uniformly across the whole length of the water line (at cathodic and anodic sites) and so sheds its paint film.

7.11 ADVANTAGES OF PRECOATED STEELS

The advantage of the coated steel is that the zinc is known to be present before the parts are stamped out. It is favoured by the industry because the waxes could be missing in corrosion prone areas and the spraying of the waxes is rather difficult to get into the areas where they are required (plate 19). The electroplated zinc has an added advantage of being sacrificial in nature and therefore provides a safety margin where there is little or no paint, eg at interwelds, (figure 3).

7.12 ZINC AND DESIGN

The presence of the zinc is ideal, if the design is correct. ***99.9% of corrosion protection is in design***. If the design is incorrect, then no matter how good the coating and how costly the materials, the corrosion process will win. One only needs a small 'corrosion leakage current' produced by perpetual anodic and cathodic sites to flow over a long time, (and 6 years is 52,560 hours), and you have pitting corrosion (figures 6(e) and 7). Therefore a vehicle parked outside with the following design defects to the doors: plates 13, 14, 15 and 16; will suffer a warranty claim and the conditions discussed in (8.4) will be applied rigorously. It is very probable that the very first paint defect will be described as stone chipping or a car park knock.

7.13 SUMMARY

The metal zinc is a good marketing tool for the manufacturers and forms the basis of the 6/8/10 year corrosion warranty. It has proven acceptability in other galvanised products as 'galvanisation'. Zinc is a good coating for steel because it corrodes in preference to steel, thus protecting steel. It corrodes giving colourless by-products (and therefore can not be seen to be corroding). Unlike the application of waxes, the manufacturers can see the presence of zinc on the steel surface prior to any body pressing (plate 19). Zinc coatings should not replace wax injection systems which is what some manufacturers are doing. Crevices still need protecting by waxes. Zinc has limitations to protection when the steel and zinc are not connected by the ***same*** liquid environment. The zinc coatings are between 7.5 and 25 microns thick.

CHAPTER 8

CORROSION WARRANTIES

8.0 TYPES OF CORROSION WARRANTIES

The **after market** rustproofing companies (Endrust, Ziebart, Dinol, Protectol, etc) had some success with the injection of waxes into internal box members. If the vehicles of that period in question had been better designed (ie been correct to reduce corrosion), these companies would have had better results. The present manufacturer warranty system owes it origination to the Canadian Government. Canada was a big importer of cars and the Canadian Government produced a ***'voluntary anti-corrosion code'*** which stated that all imports should carry a manufacturer's warranty for:

> **8 years on structural members (3.0);**
> **6 years on cosmetic panels (3.2) and**
> **18 months for the external paintwork (8.6).**

This meant that those manufacturers, wishing to export to Canada, had to take over the wax injection and warranty programme previously offered by the ***after market*** companies. It is up to the owner to detect perforation as early as possible so that corrosion is not blatantly obvious after the expiry period. These time limits are not over generous as most corrosion (eg within interwelds plate 8), can be well advanced before showing up externally. Corrosion within the interwelds reduces the vehicle's performance in a crash (3.5).

The warranties of today can be subdivided into those given by the ***manufacturers*** and those still given by the ***after market companies*** who supply the warranties for predominantly Eastern European car makers (8.12).

8.1 THE MANUFACTURER'S CORROSION WARRANTY

The corrosion warranty has now become a marketing tool. Many manufacturers will increase the length of term of the warranty within the next decade. Some manufacturers will be able to justify this increase by using better 'anti-corrosion designs' coupled with more corrosion resistant materials, but other manufacturers will just increase the period without there being any worth-while increase in the product's life. In these cases the phrase ***"I'm sorry Sir, it's stone chipping"*** will be the best you will get out of their agents.

8.2 WHAT IS EXCLUDED FROM THE WARRANTY?

The 6 year warranty is not a paint warranty, nor does it relieve the owner from touching up "wear and tear" blemishes as a result of usage. The best warranties have no time limit if the corrosion is due to a manufacturing or design fault, (8.12), conditions [19] and [22]. In others, it is a warranty to replace panels which suffer corrosion rust through from "out of sight" areas, but you will need to be a good negotiator to argue that it is not due to stone chipping. Usually the terms are interpreted as corrosion initiated on the inside and manifesting itself on the outside. ***Excluded items*** are the exhaust system, suspension components, (and they should not be excluded as corrosion in these items is due to design faults (5.11, 5.14, 5.20, 5.22, figure 23)), petrol tank, brake pipes* and in some cases external seam corrosion (3.6, 10.19). The warranty on external seams (plate 8) is really a ***must*** from the view point of keeping your vehicle in a presentable condition. Airborne fall out is excluded from some warranties, (and this may include both road de-icing salts and sea salts), but the fact that you live by the coast (figure 1) should not reduce your rights under the warranty.

* Brake pipe corrosion is excluded. Suppliers of the material suggest that when the corrosion front extends 10 thou into the pipe, there could be rupture. Corrosion pits of this magnitude are difficult to estimate. Most corrosion will occur where the tube is bent (•). When in doubt, replace, but remember corroded brakes pipes show their presence by failing under ***'panic loads'*** (•). These panic loads can be simulated by pressing both feet onto the brake pedal with as much effort as is possible. For servo assisted brakes, the engine must be running.

8.3 WAX COATED SURFACES

Some manufacturers warranties contain the same exclusion clauses which are contained within the after market warranties. These include warranties which limit the corrosion to below the waist line, window line and areas which cannot be wax injected, (see plate 9 and figure 18 for the reason). These terms are unfair in a manufacturer's warranty.

8.4 KEEPING THE WARRANTY "ALIVE"

The corrosion warranty is rather like the extended warranty on washing machines and TV's except that you don't pay the 'insurance premium'. The premium is paid for by the ***agent*** giving the warranty, (usually the manufacturer or supplier of the waxes). Dinol Protectol UK often give their own warranty to imported vehicles. The premiums work on the number of vehicles which are subjected to the liability of a replacement part, and the prior case history. The agent requiring the cover has to pay the first £100 (say) of any claim, and the remainder of the claim is paid by the insurance company. The cost of these remainders borne by the insurance company, worked out on an annual basis, reflects the insurance premium paid by the after market rustproofing company. Thus there are two factors at play here (1) keep the claim to below £100, and (2) there is a deterrent aspect whereby your vehicle is required ***'out of the system'***. These factors are achieved by claiming that the damage is due to stone chipping (and providing a goodwill gesture of 50% towards the costs), whilst the second is usually achieved by the annual inspection for which the owner pays, or by certain conditions on a change of ownership (8.11).

In contrast to this, the manufacturers carry their own insurance premium, and the inspections then become part of the mechanical servicing so that the owner takes the vehicle back to the dealership. Most of these inspections are free when carried out with the mechanical servicing thereby giving the appointed dealers a ready-made repeat trade. You will have to pay for the inspections when there is no mechanical servicing requested (approximately one hour's labour). You should be presented with a list of 'faults' which require rectification (2.7, 8.6). Don't pay anything unless you obtain this itemised fault list. The total costs of any claim are kept down by the two mechanisms discussed above.

In summary it is rather like car insurance, if you are a reckless driver and have several claims, the next year's annual premium is higher. If you don't renew your insurance and have an accident, you have no claim on your car insurance. But you could always reinsure your car if you let the insurance lapse, and this is where the two insurances differ; once you have let your 6 year warranty 'insurance' lapse, (eg by missing the inspections), you can ***not*** 'reinsure' and get back into the system. This applies to both warranty systems (manufacturer and after market warranties).

You should not therefore pay the full 'book' price if the corrosion warranty is not in force. Similarly if the vehicle has been involved in an accident (8.5) and not repaired by the manufacturer's dealership, the warranty will be void. This is most important as the corrosion warranty is not 'worth' anything to a new purchaser who keeps the vehicle for 2-3 years. The value of the corrosion warranty is in its last year, see (8.12), conditions [11] & [17]. The summary section gives an idea of where to look for the onset of corrosion. This needs to be undertaken before the warranty expires.

8.5 REPAIRS TO ACCIDENT DAMAGE

Alongside this inspection, you will also have to replace any accident damaged panels by the manufacturer's own panels and use the manufacturers materials.

8.6 EXTERNAL PAINT WORK

You will also have to keep the external paint work in good condition so that corrosion is not surface nucleated from the outside through to the inside, (figures 9 and 10). In the case of ***manufacturers warranties only***, you should get all the paint work damage* rectified within the first† months as thereafter it is excluded from the ***voluntary code***. You would be well advised to polish the vehicle and record, with the agents, any unevenness in the paint film which could be the onset of corrosion beneath the film, ie see plate 1. Insist upon stone chips on poorly designed bonnets being resprayed (10.22). This will make the manufacturers design bonnets which do not present a large flat painted area to the preceding vehicle (8.10(3)). These points will ***not*** apply to the after market warranties as they have no responsibility for these bad designs.

* Crevice interweld corrosion at the yearly examination within the 6 year warranty. You should not be charged for this rectification. For correct rectification see (10.19).

† Some manufacturers only give a 12 month warranty, whilst others give a 3 year paint warranty. More manufacturers will follow the lead of longer warranties which are used as a marketing tool to sell the cars.

8.7 CLEANLINESS

You will also have to clean the vehicle regularly. The owners handbook is not very clear on this aspect and so this booklet will attempt to show those areas where mud accumulations will start corrosion and where a layman would not expect mud to reside (eg. figure 14b, 5.5(e), 5.17).

8.8 ADDITIONS VIOLATE THE WARRANTY

In some cases, you will not be allowed to add to the manufacturers' treatment, and this will prove very difficult for authorship of a book like this.

8.9 CORROSION AND DESIGN

Corrosion is design orientated, and many vehicles are not designed to resist corrosion, in fact they are designed (by accident rather than intention) to encourage corrosion. In those cases where you are not allowed to add to the manufacturer's treatment, or make any alterations to the vehicle, I would like to make the following statement: I cannot assume, in this respect, liability or responsibility of any kind in connection with the information contained in this title which is used without first securing competent advice as to the effect upon the warranty. Therefore, it is a condition of using this booklet that any alteration to the design must be on the understanding that:

a) you are responsible initially for the alterations and their execution to high standards;

b) where the alterations are likely to materially alter the vehicle's environment, and therefore the terms and conditions of the corrosion warranty, these alterations must be authorised by the manufacturer in writing. Most manufacturers will agree to many of the suggestions contained within this book and if you are purchasing a new car, their dealers will carry out the alterations for you;

c) in the unlikely event of a manufacturer claiming that their design had been altered in such a way that the warranty is null and void, that you will assume all responsibility and appoint a solicitor to act on your behalf to sue the manufacturer. I cannot be held responsible for the application of concepts to other vehicles, nor can I possibly examine every vehicle to which these principles will/may be applied during the print run. The principles contained in this booklet and in ***Corrosion Prediction and Prevention in the Motor Vehicle***, are a distillation of my professional experience in the field of automotive corrosion.

8.10 WHY DON'T THE MANUFACTURERS MAKE THE ALTERATIONS?

The cynic would reply that historically the manufacturers have wanted their vehicles to have a limited life expectancy, it was good for their repeat business. I believe, having researched the corrosion mechanism and talked to many designers that the following are at fault:

1) many of the manufacturer's accelerated corrosion tests were wrong (•) and this gave false information on longevity;

2) the manufacturers had no unified theory as to how the vehicle corroded;

3) the designers have never considered the preceding vehicle wheel spray;

4) the designers have never considered the pressure variations which surround moving vehicles on the highway;

5) the designers have never considered the pressure eddies developed when one vehicle overtakes another or pass in opposite directions;

6) vehicle manufacturers have always tried to overcome corrosion by ventilating. This study will show that ventilation in some areas (eg, plates 3-5, 6, 7, 10, 11, 12, and 18) is detrimental to corrosion as the '**open**' holes allow further ingress of the road de-icing salts;

7) corrosion is complicated and takes in many branches of science. This is perhaps the biggest reason why the manufacturers do not get it right.

8.11 TRANSFERABILITY OF THE CORROSION WARRANTY

Most corrosion warranties now apply to the vehicle in question and not to the original owner. You should check on the conditions written into your warranty which are required on change of ownership. Some warranties require notification of change of ownership within a specified period of the sale.

8.12 WARRANTIES BY VEHICLE MAKE *(see Table A overleaf)*

8.13 SUMMARY

The 6 year warranty is not a paint warranty and does not cover the ***surface nucleated*** items of external paint work damage. The ***manufacturer's warranty*** will however cover items where corrosion has been ***designed into the vehicle*** (for example: (a) see (5.10); (b) by virtue of the constant flow of water over the painted surface (plate 1); (c) corrosion around clean rear wheel arches (figure 18, plates 8 and 9; (d) water traps behind the 'A' pillar (5.17); (e) crevice corrosion within metal overlaps etc.

The after market warranties will not cover ***any design or manufacturing fault*** and excludes areas where the wax can not reach.

The value of both warranties is in the last year before expiry.

Table A. – Warranties by Vehicle Make

Manufacturer	*No of years*	*Inspections*	*Payment required*
*Alfa Romeo	6 years	Yes, annually	Yes
Aston Martin	?	?	?
*Audi	10 years	No	Not applicable
*BMW	6 years	Yes, annually	Yes
*Citroen	6 years	Yes, annually	Yes
Dacia	6 years	Yes, annually	Yes
Daihatsu [1]	8 years	Yes, annually	Yes [1]
*Daihatsu	6 years	Yes, annually	No
Ferrari	12 months	[5b] Yes	Yes, annually
*Fiat	6 years	Yes, [4]	No
*Ford	6 years	No	Not applicable
FSO	12 months [9]		
*Honda	6 years [22]	Yes, annually	No
Hyundai [2][11]	6 years	Yes, [6a]	Yes
Isuzi	6 years [6b]	Yes	Yes [11 13 16 17 18]
*Jaguar	[19]	No	No
Lada [2][11]	6 years	Yes, [6a]	Yes [8]
*Lancia	6 years	Yes, [6a]	Yes
*Land Rover	[19]	No	No
Lotus	8 years (chassis). Panels are made from the plastic Kevlar		
*Mazda	6 years	Yes, annually	Yes
*Mercedes	12 months [7]	[6a]	[7]
*Mitsubishi	6 years	Yes, annually	Yes
*Nissan	6 years	No	Not applicable
*Opel	6 years	No	Not applicable
*Peugeot	6 years	Yes, [5]	Yes
*Porsche	10 years	No [15]	Yes, [13][14]
Proton	6 years	Yes, [10]	Yes ?
Reliant	?	?	?
*Renault	6 years	[3], Yes,	Yes.
*Rolls Royce	3 years Paint	[20]	No
*Rover	6 years	Yes, annually	No [3]
*Saab	6 years	Yes, annually	No
Seat	6 years	Yes, annually	Yes
Skoda	2 years	Not applicable	No
Subaru	6 years [6a]	Yes	Yes [11 13 16 17 18]
Suzuki	6 years	Yes [2]	Yes
*Suzuki	[?] 12 months	No. Applies to present manufacture.	
*Toyota	6 years [21]	Yes, annually	Yes
*Vauxhall	6 years	No	Not applicable
*VW	6 years	No	Not applicable
*Volvo	Lifetime	No	Not applicable
Yugo	2 years	Yes	After first year.

* = Manufacturers warranty as opposed to the importers after market warranty.

CONDITIONS

[1] Retreatments are required and charged for at the end of the 2nd and 5th years.

[2] Retreatments are required and charged for at the end of 1st, 2rd 5th years.

[3] No cost if combined with the normal mechanical service.

[4] At the end of every 12 months or 12,000 miles.

[5] At the end of every [5a] 24 and 48 months, [5b] recommended 24, 48 and 72 months, as a service item.

[6] At the end of every [6a] 12, 36 and 60 months. [6b] 12, 24, 36, 48, 60 months

[7] 12 months but the company will examine any paint/body complaint.

[8] Charge at the discretion of the dealer

[9] Manufacturer does not give a warranty. The warranty is given by the dealer as an after market warranty. These have different clauses.

[10] Requires inspection and retreatment at 18, 36 and 54 months.

[11] Warranty actually stipulates that if these services are not carried out within +/-30 days of these dates, the warranty becomes invalidated from that time and ***cannot be reinstated*** by a ***subsequent service*** see (8.4).

[13] Inspection required every year after the first two years. Inspection is free, but it is recommended that they coincide with the major service.

[14] No retreatments are required, but any damage to the underbody protection must be rectified immediately.

[15] Maintenance record is stamped to validate inspection.

[16] Claim must be made within 30 days of the rust damage becoming apparent. After this time period warranty excludes any ***additional*** rust damage.

[17] The maximum aggregate liability in respect of ***all claims*** shall not exceed 2/3 of the ***trade value*** as specified in Glasses guide at the time of repair. This gives you a rough guide to the value of a lapsed warranty.

[18] Once the payment of the maximum claim has been made, the warranty is ***null and void***.

[19] No time limit if the corrosion is due to a manufacturing fault.

[20] Manufacturers will look favourably on penetration up to 10 years provided paint rectification is carried out by an ***authorised repairer***.

[21] 3 year external cosmetic paint warranty, (stone chipping excluded).

[22] Corrosion warranty covers anything which is ***out of the control of the customer***.

[?] Information requested. No information supplied. Where information is given, this has been obtained from the information supplied with the vehicle.

CHAPTER 9

MOT and METAL PERFORATION

9.0 TEST FOR METAL PERFORATION

The problem which you, and the manufacturers, are going to face is to distinguish between surface initiated corrosion (figure 9) as a result of stone chipping, and corrosion due to 'metal perforation' from the inside to the outside (figures 7 and 25). This type of perforation corrosion is termed 'pitting'. A method of showing up pitting corrosion ***on newly formed blisters***, (denoted as L in plate 1), is shown as a cross section through the metal panel (figure 25).

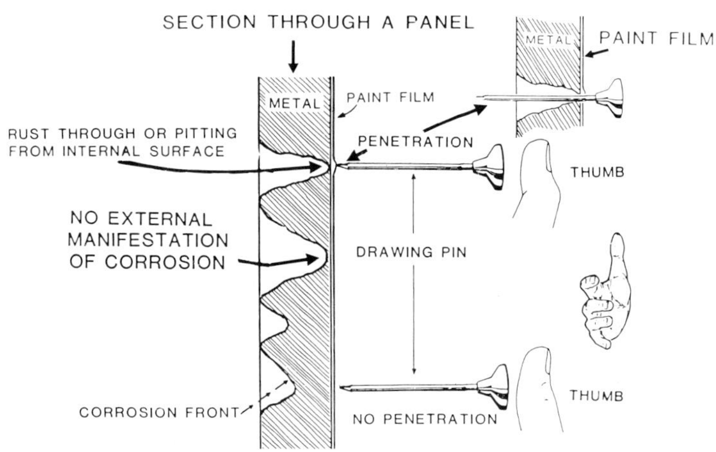

FIGURE 25. A method of showing up pitting corrosion is shown as a cross section through the metal panel. A jeweller's screwdriver, or drawing pin (something with a large flat head to spread the load on your hand), is placed upon the suspect paint blister. If the blister is due to metal perforation by pitting, then the high stresses at the tip will puncture the rust. Rust is a brittle oxide, and so has little or no resistance to high stresses which are produced by the point of the screwdriver. Metal panels of the gauge used in motor vehicle production will not be punctured by this process however hard the screwdriver is pressed. Don't be afraid of doing this test. Refer to figure 24 for the correct orientation of the pin where double skins, clinch folds, interwelds etc are suspected. If you are afraid of the screwdriver slipping, use the sharp point to pick out the rust. You will not damage the non adherent (blistered) paint surface, (figures 9 and 10), in fact you will be doing the steel surface a good turn, (see 2.4 to 2.7 for the effect which non adherent paint films have upon corrosion). The diagram also shows that a repair from the outside by 'filling in' will be short lived as other corrosion fronts, (which are not affected by the external repair), penetrate the outer surface. The effect of corrosion upon the vehicle structure and its energy absorbing properties were discussed in (3.5).

9.1 THE MOT TEST AND CORROSION

Many motorists, buying second-hand, look upon the MOT certificate as something which will reflect the vehicle's state of corrosion and its roadworthiness. A fairly high proportion of people buy a vehicle with a full 12 months MOT certificate, only to find that another expert, and/or a trading standards officer, condemns it is a "rust-box and absolutely worthless". The MOT test is a purely ***visual*** test upon the safety aspects of the car. The testers cannot remove any item, (eg road wheels, or carpets from the boot*), in order to carry out the test. This often means that the brake shoes can be worn out within the following month. They cannot damage or remove any material from the underbody. As a prospective purchaser you will only see the cosmetic panels (3.2). The MOT test does not concern itself with the corrosion of these cosmetic panels. The MOT is only concerned with the force transmitting members (shaded in figure 11d with the exception of the roof gutter and A pillar). These areas under examination are termed prescribed areas, and they are all beneath the exterior panels. Therefore inspection on a ramp is an absolute must. (More than 90% of vehicle purchases are made without the vehicle being examined on a ramp). (9.4 shows the cheapest way of hiring your own hoist).

* These ground rules may change in the future to allow removal of the carpet in the boot so that rear suspension mountings can be checked.

9.2 BEWARE OF THE CURRENT MOT

Many of the lay public put their trust in the vehicle inspectorate and they have the right to expect that this certificate should reflect vehicle road worthiness for at least 6 months from the date of test. It is worth emphasising that this is not so with the MOT in its current form. Therefore as a prospective owner of a second hand car, beware of the current MOT on the vehicle which you are thinking of purchasing, (see 9.4).

9.3 EXPIRY TIME OF THE MOT

The MOT certificate only reflects the state of the vehicle at that point in time (when it was examined). It therefore expires as you effectively walk out of the station, (except that the vehicle can be kept on the road for 12 months). With regard to body corrosion, which is only time dependent and not mileage dependent, this should not be the case. For example, in the air transport industry, you (the travelling public in general) expect the air worthiness certificates to be current for the period of time between inspections, and I believe this expectation should apply to some extent to the motor car as regards corrosion and the MOT certificate, but it does not. For tyres, and other mechanical parts which are affected by wear and tear or accidental damage this limited liability period has to hold true.

9.4 GOOD ADVICE- Support the independent MOT stations.

Having been critical of the MOT test, I am now going to contradict myself. The best advice to anyone buying a second-hand car is to get your own independent MOT at one of the independent stations which only carry out MOT's. Usually at these stations, you will be shown under the vehicle with the accredited engineer (or nominated tester). This way the hire charge of a hoist will include an engineer with a lot of experience. Explain to the tester when you fix up the appointment, that you are considering the purchase of the car and would like ***the whole car***, including the cosmetic panels examined for corrosion. There are several 'extra points' of information, (eg state of the engine, clutch wear etc), which could be passed on to you, by the examiner, because you may become one of their valued ***'once a year'*** customers. If the MOT test is carried out this way, it is a better check on the vehicle than other more expensive examinations, because at the end of the day, the MOT Certificate is a ***legal document*** on the points which it examines. Don't allow the vendor to arrange this independent test for you.

Remember: The corrosion part of the MOT only refers to structural members and for the vehicle to fail, there has to be corrosion within 12 inches (30 cm) of a "force transmitting member", (usually a suspension point attachment or seat belt anchorage). The MOT will ***not include the state of the engine, clutch, how much wear is left on the brake pads, etc.*** Under new legislation it will cover the state of all running gear, bearings and rear suspension mounts. The more expensive type of test does not always cover these points either.

9.5 PREDICTED AREAS FOR CORROSION

I have summarised in chapter 10, the areas where you should concentrate your examination when deciding on a car to buy. I have tried to explain corrosion in the simplest terms and explain the principles of how the vehicle panels perforate from rust forming reactions, which start on the internal panels, completely out of sight of everyone. I have tried to explain how corrosion is design dependent, and in so doing, I have used modern cars carrying a 6 year warranty to illustrate this corrosion. I have also tried to give some of the agents counter claims so that you can be fore warned as to what to expect. Seat belts will do damage to both surfaces (plate 15), stone chipping is random (plate 1). These principles apply to all cars. Please don't make the mistake and believe that you should understand this at the first reading. Your vehicle may not be illustrated here, but the underlying mechanism of corrosion protection associated with good design is universal.

9.6 INSPECTION PROCESS FOR CORROSION

Have the vehicle steam cleaned, or ***preferably*** washed gently with a fine spray, (see 5.6). Using a high intensity hand held torch and applying the knowledge contained in this booklet, examine the underbody, remembering to look from the front to the rear and vice versa. Look at all panels and study each panel and its drainage holes with due consideration to the vehicle direction and spray, both of the vehicle itself and those vehicles in front and overtaking. The fact that this is not undertaken during manufacture is really why we have the 'bad' designs we have. Using a plastic probe/feeler, determine whether there are any drainage hole blockages by self compacting dirt producing mud poultices in all items including suspension items. The items to examine are included in the summary.

9.7 COSTS OF CORROSION DAMAGE

At the end of the day, is all this hassle worth it? You will still want a car. Look at the problem from another viewpoint that corrosion is expensive, but how expensive is it? If you buy your own cars, you will find that a lifetime of motoring will cost you more than a roof over your head.

If you earn £100.00, you will have to pay 25% *(40%)* of it as income tax.

How much do you have to earn to pay for your £100 of corrosion damage?

Worked out at current rate of 25% income tax, the calculations are as follows:

Therefore to have £75 to spend on a car, you have to earn £100.

Therefore to have £100 to spend on a car, you will have to earn £100/75 = £133.33.

So if you have £100 worth of corrosion damage, you will have to pay £33.33 to the Tax man (as income tax) and another 15% ***(£15.00)*** to the VAT man. The gross pay required to pay the 15% VAT is £19.95.

Amount of gross income required is therefore £153.28 at 25% income tax rates.

The figures for (40%) income tax payers is:

Therefore to have £60 to spend on a car, you have to earn £100.

Therefore to have £100 to spend on a car, you will have to earn £100/60 = £166.67.

So if you have £100 worth of corrosion damage, you will have to pay £66.76 to the Tax man (as income tax) and another 15% ***(£15.00)*** to the VAT man. The gross pay required to pay the 15% VAT is £25.00.

Amount of gross income required is therefore £191.67 at 40% income tax rates.

It costs the manufacturer £100.

Remember that:

GOOD DESIGN IS FREE, BAD DESIGN IS EXPENSIVE.

It costs no more for the manufacturers to give the customer a "good design" as required for anti-corrosion measures as it does to produce a vehicle using "bad designs", (eg plates, 2, 4, 5, 6, 7, 8, 9, 10, 11, 13, 14, 15, 16, 17, 18, figures 14, 18, 21, 22). Bad design is expensive, because it is completely unnecessary and results in a waste of resources. However it is not the first owner who necessarily pays the costs of bad design. It is the second, third, and subsequent owners who pay the high price for the bad design. This is particularly true if the warranty becomes void (8.4), or if the manufacturer's goodwill gesture is accepted (10.20 to 10.22). The best warranty will look favourably on penetration up to 8 years, and will cover anything which is ***out of the control of the customer***, who may be the first, second, or third owner. If this type of warranty does not apply to your vehicle, it is important that the 'defect' (eg point L in plate 1) should be detected as early as possible, and the reason for the defect should be clarified early, (see also 10.19 to 10.22).

Chapter 10 will take the cars as they are and I will advise you what to do with the various problems which arise.

CHAPTER 10

POINTS TO REMEMBER WHEN BUYING A VEHICLE

10.1 SUMMARY AND CONCLUSIONS

Note. The following summary applies to all vehicles (cars, and commercial vehicles) built from 1950 to the present day where the form of construction is an all welded structure, (Chapter 4).

10.2 LATENT NATURAL BATTERIES IN CORROSION

Corrosion has been described as an unwanted, ***unexpected*** natural battery. A square metre of moistened steel contains many (thousands) potential corrosion sites or batteries. These latent batteries become ***active*** when the fine balance between which areas will act like the (-ve) and (+ve) electrodes of a commercial battery becomes finally established. In this book I shall only consider one external agency (figure 5) out of many (●) which establishes active sites. Natural corrosion sites are those regions which provide ***perpetually*** different oxygen replenishment rates upon a wetted metal surface ***at the same locations***. These natural batteries are: stationary water droplets (figures 1, and 6), unequal paint thicknesses (figure 3), or even at the boundary between adherent, non adherent paint surfaces (figures 9 and 10), crevices (figure 2 and 3), metal overlaps, spot welded seams (figure 2) and residual dirt (figure 6). These are all 'crevices', or in the case of dirt particles, produce crevices at the point of contact, and naturally provide an area where water can accumulate by condensation or capillary 'wicking' (figure 21). Once a water environment is present in these 'crevices', it takes a long time to evaporate and dry.

Just as there are two electrodes (+ve) and (−ve) to a battery, there are two corrosion reactions taking place at two sites called ***anodes*** and ***cathodes***. The anodes are equivalent to the (−ve) electrode of the battery (figure 4). The metal dissolves at these anodic sites when starved of oxygen producing an acidic environment (figure 7). The cathodic reaction equivalent to the (+ve) electrode produces an alkaline environment and forms where there is a plentiful supply of oxygen. ***For corrosion to take place both sites must share the same liquid environment***, (figures 3 and 7). The alkaline environment damages the paint film by reducing adhesion to the metal substrate and the acidic environment assists in the formation of pits. The reaction mechanism is called the differential aeration cell (figure 5). There is ***no*** 'mechanical switch' to switch off the 'differential aeration battery' because the metal surface is host to both (+ve) and (−ve) electrodes of the battery (figures 4, 5, 6 and 7). When all the alternatives are considered, we can only **stop** the 'battery' by ***'throwing away'*** the electrolyte (1.8, 2.0).

10.3 PERMEABILITY OF PAINTS TO WATER AND OXYGEN

A paint film will allow as much water and oxygen to pass through it, as an uncoated (ie unpainted) steel sample needs to corrode (2.4, 2.6). A paint film is impermeable to road de-icing salts for a ***short period of time***. Road dust, which may contain the winter road de-icing salts, is deposited on the vertical surfaces of the internal panels (plates 3-12) upon a very thin film of electro-deposited paint (4.4). Long term stationary particles (plates 3, 7 and 11), which are '***adhesively***' bonded to the vehicle panel by the dried road de-icing salts (figure 6c), or persistent "rivers" of water (plate 1), or trapped water (plate 15), will cause ***perpetual*** anodic and cathodic sites to develop beneath the paint film which may cause premature failure of the paint film. The failure of the paint film by ***adherence*** to its substrate allows corrosion to proceed at a greater rate (figures 9 and 10). The anodic sites become ***more active*** due to the restricted oxygen replenishment as the corrosion front progresses into the metal (figure 10). Metal perforation occurs which is claimable on the 6 year warranty provided the perforation is not nucleated from an external visual panel. ***Early detection is advisable.***

DIY Rectification: Keep the vehicle clean and free of dirt which causes perpetual sites of low oxygen replenishment. Keep the vehicle crevices coated with water hating surfaces: polish, sprayed waxes or oils (see 10.18). Use zinc paint between overlaps and insert thin polyethylene washers or water pump grease between bolted and overlapping sections.

10.4 SELF COMPACTING DIRT

High speed motorway driving produces road spray which contains very little water in relationship to the amount of dirt. This dirt can rest on vertical panels because it is not 'self cleaning' or 'self draining' (5.5). The dirt is 'self compacting' (5.4).

DIY Rectification: Hose away the mud compacts at monthly intervals during winter motoring (eg 5.6).

10.5 DRAINAGE AND PROCESS HOLES

These small particles (5.3) gain ingress into the box members (plates 3, 6 and 11) through the manufacturers process ('drainage') holes (4.2), and into suspension members via the air admittance and air exit holes (figure 23, (5.11)).

DIY Rectification: Make one way drainage flap valves figure 17. Fill process holes figure 23.

10.6 PRESSURE EDDIES AROUND A MOVING VEHICLE

The spray can alter direction very quickly due to 'pressure eddies' around the vehicle. These eddies take the road dirt and deposit the dirt in out of sight and unwashed ('unwashable') areas where they are capable of starting metal 'perforation' corrosion, eg:

a) over the wheel arch returns of any car, and within doors, and behind the wheel arch plastic stone guards. These areas are not subject to the MOT test regulations
b) within spring hangers (•), and collars (•) of suspension units (5.22)
c) rearward facing panels beneath the car to which subframes (•), and brakes pipes (•) are attached
d) within some sub frame sections (•) (plate 10)
e) within box members (•) which are perpendicular to the direction of motion (plate 18). These may have different air speeds across their ***process holes*** (4.2(b)) (figure 23, plate 18).
f) within sills (•) (rockers) (plates 11 and 12) and adjacent to jacking points (plate 17).

DIY Rectification: NB. The above (•) must be examined in the MOT test for metal perforation only in so far as they constitute a structural load bearing member within 12 inches of the metal penetration. Should you find any evidence of metal penetration (however small), reject the purchase of the vehicle, until replaced by new parts under the corrosion warranty. See figure 25 for the effects of repairs from the outside visual surface only.

10.7 MECHANISMS OF CORROSION INITIATION

There are basically eight mechanisms by which the micro-climate builds up to become a free rider on panels which are out of sight. These are:

10.7.1: COLLECTING PRECEDING VEHICLE'S SPRAY

Design defects on bonnets are shown up by any of the following: (1) The preceding vehicle spray is deflected by the radiator, headlights, rubbers or anti-vibration pads into the bonnet stiffeners. (2) See plate 7, figure 19. (3) Some bonnets become full of rain water when the vehicle is parked on uneven ground due to the orientation of the heater intake with respect to the bonnet stiffeners. (4) See also 5.9, 5.11 and 5.17.

DIY Rectification: (1) Make one way drainage flap valves figure 17. (2) Wash and keep warranty alive. (3) Park another way.

10.7.2: COLLECTING THE VEHICLE'S OWN SPRAY

Open process holes (plate 4) or 'open' seams (plate 7) can collect road spray, either from the vehicle's own front (or rear) wheel road spray, or when travelling in the preceding vehicle's road spray (figures 14a & 19).

Some panels are incorrectly designed so that they act as a 'scoop' for the vehicle's own front wheel spray (eg plate 18). See also (5.9, 5.11).

DIY Rectification: Make one way drainage flap valves and fill crevices with wax/mastic (figure 17).

10.7.3: EFFECT OF PASSENGER AIR EXTRACTION

Stale passenger air (5.12) is continually extracted through vents located in various parts of the vehicle (rear tail-gate, rear body or doors) (figure 14a). The amount of air extracted cannot be controlled at all and is dependent only upon the vehicle speed. Hence if the admittance of fresh air is reduced at high speeds, the extraction system will still operate as efficiently and lower pressures will be found within the passenger compartment and internal body cavity (box) members. (This air continuity path (plate 5) is a requirement of the mass production painting processes as discussed earlier 4.1 and 4.10). This means that replacement air (which includes low momentum road dirt (5.3) and de-icing salts) is drawn through the sills/rockers as shown in plates 11 and 12. The low momentum deposits are deposited over the rear quarter panels (plate 3), and the bottom of the doors (plate 13), from where it cannot drain readily due to badly constructed drainage holes (plate 14) and their location (plate 15).

DIY Rectification: Make one way drainage flap valves figure 17 and fit to drainage holes over the front wheel track.

10.7.4: PRESSURE EDDIES AROUND THE MOVING VEHICLE

The pressure eddies produced as one vehicle overtakes another, (or in contra-flow traffic) forces air from the high pressure side to the low pressure side through the air continuity path which exists as a requirement for mass production painting techniques. This mechanism is depicted in figure 19.

DIY Rectification: Make one way drainage flap valves figure 17 and fit to drainage holes over the front wheel track. Fit mud flaps (10.14).

10.7.5: TRAJECTORY OF WHEEL SPRAY

The wheel spray from the top of the rotating wheel is thrown forwards at twice the vehicle speed. The spray hits components of the underbody which are in front of the wheel, and facing rearward, at the vehicle speed (figure 14 a and b), and may enter box members from where they cannot be removed (plate 6).

DIY Rectification: Make one way drainage flap valves figure 17 and fit to drainage holes over the front and rear wheel track.

10.7.6: HUMID AIR, RAIN, FOG AND WIND

Entry of wind driven mist and fog through the passenger air extraction vents, or defective glazing rubbers, when the vehicle is garaged outside.

DIY Rectification: To overcome this problem, locate and detach the passenger air extractor mechanism and insert a one way valve in the extractor mechanism (figure 17). There is no easy remedy for some design defects, eg badly designed glazing rubbers, (see 5.16).

10.7.7: VEHICLES GARAGED ON SLOPES

Refer to (5.16 and 5.17).

10.7.8: SELF COMPACTING, NON DRAINING DIRT

Mud compaction takes place due to the combined effect of:

a) neutralisation of surface electric charge (•) on worn road aggregates by the road de-icing salts;

b) the high surface tension of water which leads to dense compaction of road dirt deposits in 'unwashed', and out of sight areas.

DIY Rectification: Hose away the mud compacts at monthly intervals during winter motoring.

10.8 CORROSION OF SILLS AND DOORS. See also 10.16.

The air drawn through the doors (plate 13), (which are not subjected to MOT regulations as they are considered a 'cosmetic panel'), and sills (plate 11) originates from the regions of densest spray because the process drainage holes to the doors and sills are directly over the front wheel track. This mechanism of building up the aggressive microclimate on internal (out of sight) body panels is not shown up by the manufacturer's accelerated tests undertaken on the prototype, and is discussed elsewhere (•).

DIY Rectification: Remove internal trim to door (5.16). To drainage holes in sills make one way drainage flap valves figure 17. Should you find any evidence of metal penetration (however small), reject the purchase of the vehicle, until replaced by new parts.

10.9 CORROSION DUE TO INADEQUATE UNDERBODY MAINTENANCE

In examining a vehicle for purchase, you would be well advised to ask the vendor to allow you to inspect it on a hoist. (Ask your independent MOT station as the 1/2 hour hoist charge will come with a nominated tester). Examine the vehicle for non draining dirt (5.5) in the following items; sills, suspension units and their attachments, wishbones, and all members which are load carrying members. Use a knitting needle, or a packing case strap as a 'probe' to ascertain whether the 'process holes', (4.2, 5.14, figure 23), are blocked by non draining dirt. If the process holes are blocked, remove the external dirt to the member(s) under scrutiny and examine for paint blisters and rust penetration. Test for metal perforation (9.0).

10.10 RECOGNISING THE FIRST SIGNS OF PERFORATION

On the outer visible panel(s)/item(s) metal perforation is usually shown first by a roughness which appears like dust impregnated into the paint. This is followed by a rust weeping paint blister. Later on this blister becomes a rust excrescence. At this stage it is no bigger than a pin head. The main characteristic of metal perforation is that the rust substrate cannot withstand point loading as applied by a drawing pin or jewellers screwdriver. This is the time to have the panel tested and replaced under the corrosion warranty. This test is best carried out at the dealers who act as agents for the warranty claims. Each stage should be photographed (10.22).

10.11 APPLYING THE DIFFERENTIAL AERATION THEORY

The differential aeration cell can be applied to imperceptible amounts of water in crevices formed at the base of road dirt, and interwelds (figure 3), or two body panels forming a crevice at the overlap. The region corroded is the most denuded of oxygen, (figure 7). The adjacent panel will reinforce the corroding panel and therefore the jewellers screwdriver will have to be ***directed towards the cathodic site*** of the corrosion cell, (figure 3), and as shown by the arrow on figure 24. Corrosion within these areas will usually show a ***waviness*** between certain areas (eg between spot welds) due to the volume expansion which accompanies the change of metal to rust reaction (figures 13 and 22, plate 1 at the point L), (see 10.18, 1.10).

10.12 ACCELERATED CORROSION BY DE-ICING SALTS

The de-icing salts accelerate the corrosion reaction due to the following parameters:

a) they attract water vapour from the atmosphere and hence maintain a moist environment around the dirt particle for a longer period of time (1.6).

b) they provide a good electrolyte which is a basic requirement for the interaction of the corrosion products (1.10, 2.1, 2.5).

c) when drying out, the salt concentrates at the base of the dust particle, (figure 6c), or in the interwelds which have restricted drying abilities, (1.6).

d) the salt acts an '***adhesive***' sticking the dirt particle onto vertical surfaces, thereby producing ***perpetual anodic and cathodic sites.***

e) the areas where they concentrate are those areas which produce a 'battery effect' which arises from the differences in the replenishment rate of oxygen to the wet steel surface.

f) the corrosion cell produced by dirt particles, contaminated with road de-icing salts, has a large cathode and a very small anode which promotes pitting (figure 7).

DIY Rectification: Keep the vehicle clean and free of perpetual sites of low oxygen replenishment. Keep the vehicle crevices coated with water hating surfaces: polish, sprayed waxes or oils. Hose away the mud compacts at monthly intervals during winter motoring.

10.13 PREDICTED AREAS OF CORROSION

Predicted areas of corrosion are adjacent to open electrocoat 'drainage' holes, (5.13). The metal perforation due to corrosion often bears a 1:1 correlation with the manufacturers drainage/process holes (plates 11 and 12).

10.14 PROTUBERANCES. JACKING POINTS, MUDFLAPS, ETC.

Predicted areas of corrosion are adjacent to members which form a protuberance to the preceding vehicle spray and/or front (or rear) wheel spray. The protuberances will cause a rise in pressure and jettison the road de-icing salts into the box members (plates 7, 10, 17, and 18).

DIY Rectification: Make one way drainage flap valves (figure 17) and completely cover all crevices (or holes) which will collect the wheel spray. Fill process holes if the mudflap, suspension attachment or jacking point etc, is adjacent to any hole.

10.15 CORROSION WITHIN BOX MEMBERS

Predicted areas of corrosion are within box members which are subjected to different air speeds at their open ends as a result of their paint processing holes (4.2, plate 18). The different air speed will generate pressure differences between these "openings" which will lead to the introduction of a corrosive environment. Open process holes will collect the road spray (figure 20). See also plate 6, (5.11, 5.14).

DIY Rectification: Make one way drainage flap valves figure 17 to cover lower open end.

10.16 DESIGN OF DOORS AND EFFECT UPON CORROSION

Predicted areas of corrosion are within doors which contain badly designed door glazing rubbers (plate 16), or drainage holes which are not at either corner (plate 15), or which do not allow unimpeded exit of the rain water (figure 21(b)(C), or collect the front wheel spray (figure 21b (B)). See also (5.16, 7.12).

DIY Rectification: Keep warranty alive, examine periodically, should you find any evidence of metal penetration (however small), reject the purchase of the vehicle, until replaced by new parts. See also (5.16).

10.17 REPAIRS AND OLD VEHICLES

In general, look for areas which are rough, newly painted, or coated with mastic.

DIY Rectification: Repaired areas in stress bearing members will require examination with a magnet. If the only available magnet is too strong, use a plastic "spacer" to reduce the attractive force thereby allowing the differentiation of filler materials and steel. Aluminium and fibreglass are non magnetic materials. Reject purchase if filled by fibreglass, aluminium, pop rivets, etc. The vehicle will not get another MOT without these areas being rectified (3.3).

10.18 CORROSION WITHIN CREVICES AND INTERWELDS

Pay particular attention to crevices between spot welds (interwelds), metal overlaps, bolted sections and aerial fixing points. Run your hand along the interweld flange and determine whether they are distorted, wavy or bulging between spot welds. Distorted interweld flanges will indicate corrosion within the interweld (figure 22). This is shown up on the wing, plate 1 near the point L. The alkali produced by the corrosion reaction strips the paint off both sides of the wetted crevice.

DIY Rectification: Should you find any evidence of interweld corrosion reject the purchase of the vehicle, until replaced by new parts. Start claim on 6 year warranty (see next section). Paint stripping will occur on both surfaces (eg aerials (1.10)) and so should be repaired from both sides.

10.19 IRREPARABILITY OF INTERWELD CORROSION

Corroding spot welded crevices cannot be repaired. To correctly rectify the corrosion shown in plates 3, 6, 8, 9, 13,14, 15, 17, 18, 15, (eg due to a manufacturing fault arising from a lack of wax within the crevice as shown in plates 8 and 9), would require the complete removal of the panel to expose the crevice. This would have to be followed by a good rub down to remove the rust, followed by the application of a conversion coating (2.1), and repaired with a zinc rich interweld primer between the remade spot welds. As you know, if this were due to stone chipping, one could spray the arch from the outside and all will be well. It is impossible to repair corroded interwelds from the outside on the assumption that the corrosion is due to stone chipping.

DIY Rectification: The only method available, which will reduce the ***rate*** of corrosion (but not affect the cause), is to thoroughly dry the area and allow penetrating oil to flood the area well. This thin oil will penetrate the crevice and exclude water by virtue of its contact angle (figure 21a, substrate B). The ***thickness*** of the oil does not play a part in the protective mechanism. As oils are easily washed away, this crevice area should be protected externally by: *cellulose paint, mud flaps, (water pump) grease film etc. depending upon the aesthetics of the area in question. (* remove oil first).

10.20 MAKING A CLAIM. THE 50% GOODWILL GESTURE

Ensure that the warranty is ***alive*** and you understand the warranty conditions. Early identification coupled with an understanding of the cause and the principles behind the defect is important. You will be certainly offered a goodwill gesture rectification of the damage where the manufacturer may offer to pay 50%. Correct identification of the problem is important as the other 50% will have to be paid by yourself (9.7), and it may be that this 50% from you, is keeping the total warranty costs down (8.4). A good example of this is the rear wheel arch, (plate 8, and repair (10.19)). Stones only travel in straight lines, and a lot of the rear wheel arch damage will be ***wrongly*** ascribed to ***stone chipping****. The majority of the damage here will be due to the loss of paint adhesion resulting from the alkali produced as a by-product from the cathodic reaction (1.10). The actual loss of metal at the anodic site will be occurring in the crevice ***completely out of sight***), probably fed by the mud pocket which will rest on this lip. If you have washed under the rear wheel arches, you should not accept the manufacturer's gesture of goodwill as you will be in greater trouble later on. Remember that overtaking vehicles will cause side impact damage, but this will be more noticeable on the doors. Check for these door blemishes. If there are none, then it is hardly likely that the other damage is due to stone chipping. *See also: (5.7); plate 6; (5.9); figure 14b; plate 18; (7.12).

10.21 REMEMBER: Correct identification of cause is important

If you take up some of the goodwill gestures which are offered to you, you will be actually paying for something which may be shown to be ***wholly*** the manufacturer's responsibility, and furthermore and perhaps more important, the ***cause and source of the rust will still be present***. When the defect reappears you will have no grounds for redress as the vehicle may now be out of the 6 year warranty period, so be warned. A saying which summarises all motoring problems is: ***If you can't make it right, make it shine.*** Apply this to all (new and) second hand cars and you won't go far wrong. If the rear wheel arches on a second hand car shine like a new pin, examine them carefully using a mirror (plates 8 and 9) because they were probably resprayed ***due to stone chipping***. I hope you will now appreciate that they were resprayed due to crevice corrosion and that this repair is short lived.

10.22 APPOINTING LEGAL REPRESENTATIVES

If you do appoint a solicitor to act on your behalf, the contract is between you and the solicitor and if you accept a 'better goodwill gesture' when you have been arguing for 6 months, you will find that the legal representative has to be paid by you. ***Don't*** appoint a legal representative until you have had a good try at conducting your own discussions with the manufacturers. A solicitor cannot help you on the technicalities of corrosion (eg of plates 1, 6, 8, 15, 16, 17, and 18; (5.17)) and may suggest that the goodwill offer is acceptable when he has considered all the provisions of the warranty. This could be to your disadvantage. Another approach, if you feel that the vehicle is of ***unmerchantable quality***, eg plates 1, 4, 6, 9, 15, 17, or where the bonnet presents a large area which is perpendicular to the preceding vehicle spray (figure 14a), you† could take out a small claims action against the retailer. This is a fixed* fee service (*amount depending upon the size of claim) where, if you convince the court and win, your fee is recovered. Even though this is a manufacturing "fault," the retailer could be considered liable simply because your contract is made, in law with the retailer. The retailer would then have to seek redress from the manufacturer.

Before appointing legal representatives, or proceeding through the small claims court, I would suggest taking photographs of the offending areas on your vehicle ***before*** and ***after*** tests (eg plate 1 blister L is a "before" plate, whilst plate 12, if it were of the same area, would be an "after" plate showing the application of the principles shown in figure 25). ***Don't*** photograph the ***before*** and ***after*** tests on the same roll of film as sometimes the photographs are overexposed as a result of flash bounce. Flash bounce is considerably reduced if you photograph the panels at an angle. Present your case using the scientific basis for your case. If you don't get satisfaction, then present your case to the small claims court, or appoint your legal representative and then he can present your case using your material.

† Applies to first owner only.